THE

WOMAN SUFFRAGE COOK BOOK

SECOND EDITION.

THE

WOMAN SUFFRAGE COOK BOOK,

CONTAINING THOROUGHLY TESTED AND RELIABLE RECIPES FOR COOKING, DIRECTIONS FOR THE CARE OF THE SICK, AND PRACTICAL SUGGESTIONS,

CONTRIBUTED ESPECIALLY FOR THIS WORK.

EDITED AND PUBLISHED BY

MRS. HATTIE A. BURR,

12 WAYNE STREET, BOSTON.

In what thou eatest and drinkest, seeking from thence
Due nourishment, not gluttonous delight,
So mayest thou live, till, like ripe fruit, thou drop
Into thy mother's lap; or be with ease
Gathered, not harshly plucked; for death mature.

MILTON.

Facsimile Edition

APPLEWOOD BOOKS
Carlisle, Massachusetts

978-1-4290-9540-2

Thank you for purchasing an Applewood book.
Applewood reprints America's lively classics—
books from the past that are still of
interest to modern readers.
Our mission is to build a picture of America's
past through its primary sources.

To inquire about this edition
or to request a free copy
of our current catalog
featuring our best-selling books, write to:
Applewood Books
P.O. Box 27
Carlisle, MA 01741
For more complete listings,
visit us on the web at:
www.awb.com

10 9 8 7 6 5 4 3

MANUFACTURED IN THE UNITED STATES OF AMERICA

PREFACE.

This little volume is sent out with an important mission. It has been carefully prepared, and will prove a practical, reliable authority on cookery, housekeeping, and care of the sick, especially adapted to family use. While many of the receipts are original, it is not claimed that all are so; but each has been thoroughly tested, and is vouched for as reliable by the contributor whose name is appended.

Among the contributors are many who are eminent in their professions as teachers, lecturers, physicians, ministers, and authors,—whose names are household words in the land. A book with so unique and notable a list of contributors, vouched for by such undoubted authority, has never before been given to the public.

Grateful acknowledgments are due to the kind friends,—many of them in distant homes,—who have so willingly contributed of their knowledge and experience for the accomplishment of this undertaking. I believe the great value of these contributions will be fully appreciated, and our messenger will go forth a blessing to housekeepers, and an advocate for the elevation and enfranchisement of woman.

HATTIE A. BURR.

BOSTON, NOVEMBER 25, 1886.

CONTRIBUTORS.

Mary A. Livermore,	Melrose.
Lucy Stone,	Boston.
Mrs. Ednah D. Cheney,	Boston.
Mrs. Jane L. Patterson,	Boston.
Mrs. Abby Kelley Foster,	Worcester.
Mrs. Zilpha H. Spooner,	Plymouth.
Rev. Ada C. Bowles,	Abington.
Rebecca Moore,	London, Eng.
Mrs. Oliver Ames,	Boston.
Mary J. Safford, M. D.	Boston.
Abigail Scott Duniway,	Portland, O.
Mrs. H. H. Robinson,	Malden.
Mrs. Harriette R. Shattuck,	Malden.
Mrs. A. A. Miner,	Boston.
Frances Willard,	Evanston, Ill.
Mrs. J. Blackmer,	Springfield.
Sarah R. Bowditch (Mrs. William I.),	Brookline.
Marie E. Zakrzewska, M. D.	Boston.
Lillie Devereux Blake,	New York.
Clara Berwick Colby,	Beatrice, Neb.
Elizabeth L. Saxon,	Beatrice, Neb.
Anna Ella Carroll,	Washington, D. C.
Mrs. C. C. Hussey,	East Orange, N. J.
Mrs. Benj. F. Pitman,	Somerville.
Mrs. William C. Collar,	Boston.
Sarah R. May (Mrs. Samuel),	Leicester.
Mary F. Daniell (Mrs. M. Grant),	Boston.
Julia Ward Howe,	Boston.
May Wright Sewall,	Indianapolis, Ind.
Cora Scott Pond,	Boston.
Rev. Annie H. Shaw,	Boston.
Mrs. Matilda Joslyn Gage,	Fayetteville, N. Y.
Emily A. Fifield,	Boston.

Abby Morton Diaz,	Boston.
Clemence S. Lozier, M. D.	New York.
Cora L. Stockham,	Chicago, Ill.
Mrs. Mary F. Curtiss,	Boston.
Mrs. J. W. Guiteau,	New York.
Mary Gay Capen,	Boston.
Annie Jenness Miller,	Washington, D. C.
Mrs. S. H. Richards,	Weymouth.
Miss Harriet B. Hicks,	Needham.
Dr. Vesta D. Miller,	Needham.
Miss Mary Willey,	Boston.
Paul Gates,	Ashby.
Mrs. L. Clementine Gates,	Ashby.
Mrs. Marcia E. P. Hunt,	Weymouth.
Angelina Ricketson,	New Bedford.
Miss Josephine P. Holland,	Foxborough.
Mrs. Martha J. Waite,	New Bedford.
Mrs. Mary S. Tarbell,	Ayer.
Mrs. Mary C. Ames,	Boston.
Louisa G. Aldrich,	Fall River.
Mrs. C. M. Ransom,	Newton.
Miss Emma A. Ransom,	Newton.
Mrs. Elizabeth W. Stanton,	New Bedford.
Mrs. Harriet M. Turner,	Boston.
Dr. Caroline E. Hastings,	Boston.
Mrs. Emma P. Ewing,	Ames, Iowa.
Mrs. Susie C. Vogl,	Boston.
Miss Hattie E. Turner,	Boston.
Miss C. Wilde,	Boston.
Miss J. S. Foster,	Philadelphia.
Mrs. Elizabeth Groshan,	Beachmont.
Julia A. Kellogg,	Somerville.
Martha B. Pitman,	Somerville.
Mary Cross Harris,	Worcester.
Alice Stone Blackwell,	Boston.
Miss Lucy Goddard,	Boston.
Mrs. Alice A. Geddes,	Cambridge.
Mrs. S. J. G. Beck,	Freeport, Me.
Mrs. Emma E. Foster,	Worcester.
Mrs. S. J. Vineelette,	East Douglass.
Mrs. Mary J. Buchanan,	Boston.
Mrs. Sarah M. Perkins,	Duxbury.
Mrs. Alice M. Southwick,	Uxbridge.
Mrs. Forrest W. Forbes,	Westboro'.
Carrie A. Sargent,	Haverhill.
Mrs. N. W. Lyon (Mrs. Thomas S.),	Topeka, Kan.
Mrs. H. Andrews,	Fitchburg.

Rev. Louise S. Baker, ... Nantucket.
Sara T. L. Robinson, ... Lawrence, Kan.
Mrs. Mary F. Crowell, ... Dennis.
Mrs. Ruth F. Elwell, ... Boston.
Mrs. Ellie A. Hill, ... Natick.
Miss C. Wellington, ... Lexington.
Hulda B. Loud, ... Rockland.
Sarah A. Loud, ... Rockland.
Emma M. E. Sanborn, M. D., ... Andover.
Mrs. Jessie F. A. Banks (Mrs. Louis A.), . Boston.
Mrs. Ellen W. E. Parton, ... Newburyport.
Mrs. M. M. Woolford, ... Ayer.
Mrs. Mary J. Willis, ... Ayer.
Helen V. Austin, ... Johnstown, Pa.
Mrs. Allie E. Whitaker, ... Boston.
Anna B. Taylor, M. D. . . . . Charlestown.
Miss L. A. Hatch, ... Boston.
Mrs. Dr. C. A. Carlton, ... Salem.
Mrs. O. A. Cheney, ... Natick.
Mrs. M. L. T. Hidden, ... Lyndon, Vt.
Mrs. S. R. Urbino, ... Boston.
Mrs. F. D. Osgood, ... Boston.
Mrs. Josie Currier (Mrs. S. E. D.), ... Boston.
Evelyn Greenleaf Sutherland, ... Arredondo, Fla.
Catherine H. Birney, ... Washington, D. C.
Mrs. H. O. Hawkins, ... Taunton.
Mrs. L. W. Jones, ... Malden.
Miss A. E. Newell, ... Boston.
Miss L. F. S. Barnard, ... Boston.
Mrs. S. W. Fuller, ... Boston.
Dr. Leila G. Bedell, ... Chicago, Ill.
Emily S. Bouton, ... Toledo, O.
Miss M. L. Moreland, ... Fitchburg.
Mrs. Dr. Flavel S. Thomas, M. S. . . . . Hanson.
Dr. Alice M. Eaton, ... Chelsea.
Miss E. B. Plympton, ... Woburn.
Mrs. M. F. Walling, ... Cambridge.
Mrs. S. C. Lincoln, ... Somerville.
Mrs. Harriet C. Batchelder, ... S. Framingham.
Mrs. M. E. Sammett, ... Dedham.
Miss M. A. Hill, ... Haverhill.
Mrs. B. J. Stone, ... Westborough.
Mrs. E. J. Harding, ... S. Weymouth.
Rebecca Howland, ... Fairhaven.
Alice B. Stockham, M. D. . . . . . . Chicago, Ill.
Mrs Judith W. Smith, ... Boston.
Amanda M. Lougee, ... Boston.

Mrs. Martha A. Everett,	Dover.
Louise V. Boyd,	Dublin, Ind.
Mrs. D. W. Gage,	Oberlin, O.
Miss Jerusha S. Hall,	Dennis.
Mrs. H. A. Foster,	Somerville.
S. Louise Simonds,	Belmont.
Mrs. S. C. Wrightington,	Fall River.
Sarah F. Sargent,	Malden.
Ella C. Elder,	Florence.
Anne B. Rogers,	Rockport.
Mrs. M. C. Whittier,	Fitchburg.
Mrs. Richardson,	Fitchburg.
Mrs. D. B. Whittier,	Fitchburg.
Mrs. Augusta Rich,	Osterville.
Mrs. Sarah E. M. Kingsbury,	Needham.
Mrs. Mariana T. Folsom,	San Antonio, Texas.
Mrs. D. P. Washburn,	Needham.
Mrs. Elizabeth C. Crosby,	Nantucket.
Helen B. W. Worth,	Nantucket.
Mrs. Eliza B. Burgess,	Nantucket.
Linda S. Barney,	Nantucket.
Katherine Starbuck,	Nantucket.
Sarah S. Swain,	Nantucket.
S. Adelaide Hall, M. D.	Watertown.
Mrs. E. R. Abbott,	Andover.
Mrs. B. M. Nichols,	Foxborough.
Miss Harriet Lemist,	Boston.
Mrs. E. E. Kelsey,	Somerville.
Jane Hosmer,	Concord.
Jennie W. Smith,	Boston.
Jennie S. Harrison,	Plymouth.
Mrs. M. Angelo Foster,	Watertown.
Minnie I. Estes Churchill,	Brockton.
Martha Anderson,	Boston.
Mary F. Holmes,	Oxford, Me.
Mrs. Esther A. Jobes,	Billings, M. T.
Mrs. Mary H. Hunt,	Hyde Park.

CONTENTS.

INDEX.

VEGETABLES AND SIDE DISHES.

SALADS, PICKLES, &c.

PUDDINGS.

PIES.

DESSERTS, CREAMS, ETC.

CAKES.

PRESERVED AND CANNED FRUITS, JELLIES, ETC.

COOKING FOR AND CARE OF INVALIDS.

MISCELLANEOUS.

BREAD AND YEAST.

Bread.

Boil one pint or one quart of milk, according to the quantity of bread required. Pour it on the flour, and stir with a spoon until of the consistency of what our grandmothers called "popped robins." Add cold water, mixing with the hand. When cool enough not to scald the yeast, add a cup, and knead until it will not stick to the board,—about half an hour. Let it rise over night. Make into loaves or breakfast biscuit; let it rise again and bake.

MRS. JANE L. PATTERSON.

Bread.

Dissolve an ounce cake of Fleischmann's, or some other good compressed yeast, and a teaspoonful of salt, in a quart of lukewarm wetting—either milk, or water, or milk and water in equal proportion—and gradually stir in flour with a wooden spoon until the dough is of sufficient consistency to be turned or lifted from the bowl in a mass. Add flour as desired, until it can be worked without sticking to the molding board or the fingers, then put in a warm earthen bowl, well greased, cover with a bread towel and blanket, and set to rise till light, which, if kept at a temperature of 75°, will be in about three hours. As soon as sufficiently light, form into loaves or rolls, put into greased pans, cover as before, and again set to rise for an hour, or until light, and then bake. The surface of the dough should be lightly brushed with melted butter before it is set to rise, to keep it from becoming dry and hard, and the oven should be at the proper temperature when the bread is put in it, and should be kept so during the entire period of baking. If this recipe is strictly followed, and the yeast and flour are of good quality, it will invariably produce sweet, nutty-flavored, delicious bread and rolls. MRS. EMMA P. EWING.

Brown Bread.

Three cups Indian meal, three cups rye meal, one cup molasses, one teaspoonful saleratus; work up with milk about as thick as johnny-cake, butter the steamer, pour in, cook about five hours. MRS. SARAH R. BOWDITCH.

Brown Bread.

Two cups of Indian meal, two cups of rye meal, one cup of flour, one large cup of molasses, one teaspoonful of soda. Mix soft with warm water. Steam five hours.

MRS. ZILPHA H. SPOONER.

Brown Bread.

Two cups yellow corn meal, two cups sifted graham, two-thirds cup molasses, one-half cup raisins, one small teaspoon salt, one teaspoon *full* of soda; mix very soft, with butter-milk, sour milk, or cold water. Steam four hours, finish in the oven one-half hour. I prefer an earthen dish for the better cooking. A little less soda when water is used.

MRS. J. BLACKMER.

Iowa Brown Bread.

Ingredients: three cups corn meal, two cups rye meal, three cups sour milk, one cup N. O. molasses, one cup raisins, two teaspoons salt, three teaspoons soda. Process: sift the corn and rye meal together. Mix the milk, molasses and salt together. Dissolve the soda in a little warm water. Pour the dissolved soda into the milk and molasses, and, while the mixture is effervescing, pour it into the meal—beating with a wooden spoon until smooth. Grease a pudding-boiler and pour in the batter, a little at a time, adding the raisins in layers, until the mould is filled to within about two inches of the top. Cover closely, place in a kettle of boiling water and cook four or five hours, adding more boiling water as that in the kettle evaporates.

MRS. EMMA P. EWING.

Steamed Brown Bread.

One quart rye meal, a small pint Indian meal well sifted, three teaspoons Royal Baking Powder stirred thoroughly into the meal, half a cup molasses, two-thirds teaspoon soda dissolved in quite hot water with a piece of butter size of a large walnut. (The soda is for the rye meal and molasses.) Wet the mixture with warm water and milk or clear warm water. Steam in tin or earthen dish six or eight hours. It may be put into the oven half an hour or more to form a crust, if so liked. MRS. MARY S. TARBELL.

Oatmeal or Rice Bread.

Two cups cooked oat meal, or rice, salt to taste, two tablespoonfuls of sugar, one cup sweet milk, one-third cup yeast, flour to make it stiff. S. LOUISE SIMONDS.

Raised Bread.

Scald one pint of Indian meal with two quarts of *boiling* water, add as much flour as you can stir in with a spoon, let it set until cool, then add one-half yeast cake dissolved in a cup of warm water, add another cup of warm water, one tablespoonful of salt, one tablespoonful of sugar, and flour enough to knead, but do not make it as stiff as ordinary raised bread. Let it rise over night, then make it into loaves and let it rise again, then bake. MRS. M. E. SAMMET.

Pure Salt Rising Bread.

When the kitchen fire is lighted in the morning put a quart cup, one-third full of fresh water, on the range and heat it quickly to 95°. Remove from the fire, add a teaspoonful of salt, a pinch of brown sugar, and coarse flour or middlings sufficient to make a batter of about the right consistency for griddle cakes. Set the cup, with the spoon in it, in a closed vessel half filled with water moderately hot but not scalding. Keep the temperature as nearly even as possible, and add a spoonful of flour once or twice during the process of fermentation. The yeast ought to reach the top of the bowl in about five hours. Dip your flour into a tray or pan, make an opening in the centre and pour in your yeast. Have ready a pitcher of warm milk, salted, or milk and water (not too hot, or you will scald the yeast germs), and stir rapidly into a pulpy mass with a spoon. Cover this sponge closely and keep warm for an hour, then knead into loaves, adding flour to make the proper consistency. Place in warm well-greased pans, cover closely, and leave till it is light. Bake in a steady oven, and when done let all the hot steam escape. Wrap closely in damp towels and keep in closed earthen jars till wanted. There is no sweeter, nicer, better, or more wholesome bread than this; but it takes time, patience and thought to make it. Try it, and be convinced. ABIGAIL SCOTT DUNIWAY.

To Make Yeast.

Boil four large, pared potatoes in two quarts of water, and strain all through a colander. Stir two heaping tablespoonfuls of flour evenly in a quart of water, and boil it up once or twice. Pour a pint of boiling hot water over a tablespoonful of pressed hops and allow it to stand a few minutes. Now stir together the mashed potato and the liquid in which they were boiled, the flour and water which have been boiled together, and the liquid in which the hops have steeped. Add two tablespoonfuls of sugar and one of salt. Set away to cool until it is lukewarm, so that you can hold your finger in the mixture, then add a cupful of good yeast or one yeast cake. Keep the mixture moderately warm until it rises, as it will in four or six hours. Then cork it tightly in a stone jug, and put it away in the refrigerator or a cool cellar. You will have the best yeast that can be made for family use. MARY A. LIVERMORE.

Yeast.

One cup sugar, one half cup salt, pare and grate three or four good-sized potatoes, add two quarts boiling water, and let all boil together five minutes. Steep a pinch of hops in a half-pint of boiling water, and add to the yeast. Set to cool. When blood-warm add a cup of the same kind of yeast — saved from your own or borrowed of your neighbor. Let it rise, then set it in a cool cellar, or refrigerator, and it will keep until used up — a month or six weeks — in perfect sweetness. JANE L. PATTERSON.

Home Made Yeast.

Boil a heaping quart of loose hops (or if they are pressed, two ounces) in one gallon of water, strain it, when it is cold put in a small handful of salt, and a half pound of sugar, then take a pound of flour and rub it smooth with some of the liquor, after which make it thin with more of the same liquor, and mix all together, let this stand twenty-four hours; then boil and mash three pounds of potatoes and add to it, let it stand twenty-four hours more; then put it in a bottle or a tight vessel, and it is ready for use. Shake the bottle before using. It should be kept in a warm place while it is making, and in a cool place afterward.

LUCY STONE.

Good Hop Yeast.

Boil a tablespoonful of hops in a quart of water; grate two pared potatoes in a pan, add two tablespoonfuls of sugar, two of flour, one of salt; stir all together, and then add the strained yeast water; stir briskly over a hot fire till the mixture comes to a boil, then put away to cool; when nearly cold, add a teacupful of yeast; when light and foamy, put in a clean stone jar and cover tightly and keep in a cool place. This will keep several weeks, and will make the most delicious bread.

MRS. SARAH M. PERKINS.

BREAKFAST AND TEA CAKES.

Breakfast Gems.

One cup water, one cup milk, two cups flour; beat from ten to fifteen minutes rapidly with an egg-beater. These will rise like popovers. No salt, the cups of flour *not* heaped; bake in a quick oven. MARTHA B. PITMAN.

Indian Cake.

Sift fresh Indian meal and salt it, moisten with *boiling* water, beat it well; add boiling water by degrees, beating it perfectly smooth, until it is a thin batter; pour it into hot, buttered, tin or gem pans, and bake in a quick oven.

"Thin in the batter, thick in the pan," is an old cook's rule. The thickness is a matter of taste.

Indian meal prepared with boiling milk by the same process, gives a pleasing variety. A hot mashed sweet potato mixed with the hot meal is good.

LUCY GODDARD.

Bannock.

One and one-half pints Indian meal, three pints boiled milk, nine eggs, one coffee-cup sugar, salt. Bake two hours. The eggs should be beaten and ready to mix with the meal and milk while hot, and it should be baked immediately. LINDA S. BARNEY.

Squash Biscuit.

One cup sweet milk, one of sifted squash, one tablespoonful of butter, one of sugar, one egg, three cups of flour, with one teaspoonful of cream tartar and one-half of soda, and a little salt thoroughly mixed through the flour. Bake in roll iron. Have the iron hot.

ANNA B. TAYLOR, M. D.

Squash Biscuit.

One pint squash (boiled and strained), one cup sugar, one-half cup yeast, two-thirds cup butter, or one-half lard, little salt, one-half teaspoonful soda. Knead stiff. Let it rise very light before and after it is put in the pans. For breakfast, put in pans at night. MRS. RICHARDSON.

Washington Tea Biscuit.

One quart of flour, one salt-spoon of salt, small teacup of yeast, or a small piece of compressed yeast dissolved in water, one tablespoonful of sugar, one egg, a piece of lard the size of a walnut. Rub the flour and lard together, add the salt, yeast, egg and sugar, with enough warm water or milk to make a thick batter. When well risen, add more flour, make into rolls and bake in a quick oven.

MISS E. B. PLYMPTON.

Breakfast Cakes.—(Popovers.)

Three heaped cups flour, two of milk and one of water, a teaspoonful of salt. Beat well; add a beaten egg. Pour into smoking-hot roll-pans, and bake half an hour. Graham by the same method: One cup of flour, two of graham.

JANE L. PATTERSON.

Albany Breakfast Cakes.

One pint of milk, one cup of white flour, one cup of Dansville wheat-flour, one egg. Mix smoothly and bake quickly in gem pan, which should be first heated on top of the stove and allowed to remain about five minutes after filling before putting in the oven. MRS. S. H. RICHARDS.

Huckleberry Breakfast Cake.

Two eggs, one-half cup sugar, one teaspoonful of butter, one pint flour, two teaspoonfuls baking-powder, one pint berries freshly washed to add moisture to the flour, one tablespoonful water. Bake in long pie tin.

DR. LEILA G. BEDELL.

Katie's Breakfast Cakes.

Two and one-half cups of flour, two eggs, two tablespoonfuls sugar, one cup milk, salt, baking-powder.

MISS A. E. NEWELL.

Breakfast Oat-Meal.

Take, at night, a tin vessel you wish full of cooked oatmeal, and fill it little more than a third full of dry meal, wash it, put in salt, and fill the vessel nearly full of water. In the morning place this tin vessel of soaked meal in an iron pot that contains a little water. Cover the tin vessel and the iron one, and boil. In little less than an hour the tin vessel will be full of well-cooked oatmeal ready to serve.

MARIANA T. FOLSOM.

Corn Cake.

One cup corn-meal, two cups flour, five dessertspoonfuls sugar, two teaspoonfuls cream of tartar. Stir these well together. Add two eggs well beaten, one teaspoonful soda, one pint milk. DR. VESTA D. MILLER.

Corn Cake.

One cup Indian-meal, one cup flour, one-half cup sugar, one cup sweet milk, one egg, two tablespoonfuls of lard, one teaspoonful cream of tartar, one-half teaspoonful of soda. Bake in roll-iron, or in thin sheets.

ANNA B. TAYLOR, M. D.

Best Corn Cake.

One egg, one-half cup of sugar, one cup of milk, one cup of Indian-meal, one cup of flour, one teaspoonful of soda, two teaspoonfuls of cream of tartar. Bake in a loaf or in small tins, or for a variety bake it in a thin sheet in a dripping pan, and cut in squares. It will be very light.

ALLIE E. WHITTAKER.

Delicious Corn Cake.

Thoroughly mix one cup of Indian-meal, one cup flour, one spoonful of granulated sugar, a little salt, one teaspoonful of cream of tartar. Beat to a froth one egg, add one-half a teaspoonful of saleratus, dissolved in a cup of milk. Beat the whole, hard, for a minute, put quickly into a hot oven. Will bake in a few minutes.

MISS L. A. HATCH.

Doughnuts.

One-half cup cream, one-half cup sugar, three and a half cups St. Louis flour, one-half teaspoonful soda, one teaspoonful cream tartar, two eggs. Spice of any kind you like best, or nothing but one good-sized teaspoonful of salt.

MRS. D. W. FORBES.

Doughnuts.

Two eggs, one and a half cups of sour milk, a half cup of thick cream, or three tablespoonfuls of melted butter, one cup of sugar, a half-teaspoonful of soda (if the milk is very sour, a little more), and flour enough to mix soft. When butter is used instead of cream, mix the butter, sugar and eggs together first.

MRS. L. M. T. HIDDEN.

Doughnuts.

One cup of sweet milk, one half cup of sweet cream, one and one-half cups of sugar, three eggs, well beaten with Dover egg beater, one half nutmeg, one teaspoonful salt, two teaspoonfuls cream of tartar, one teaspoonful soda, flour enough to roll out, though not very stiff. Cut round, with hole in centre, and fry in hot lard.

MRS. DR. C. A. CARLTON.

Doughnuts.

One cup sugar, butter one-half size of an egg, two eggs, well beaten, one cup sour milk, one-half teaspoonful of salt, one-half teaspoonful of soda, as little flour as possible to roll out. S. LOUISE SIMONDS.

Children's Doughnuts.

One cup sweet milk, two cups sugar, three eggs, lemon flavoring, three heaping teaspoonfuls baking powder. Sift about two quarts flour into mixing pan, making place in the centre for baking powder, sugar, eggs, flavoring, and butter size of walnut. Add the milk, mixing slowly, and use enough flour to roll without sticking. Roll quite thin; cut in rings, and fry in smoking hot lard. Drain well. Equal parts of lard and beef fat may be used.

MRS. JESSIE F. A. BANKS.

Breakfast Gems.

One pint milk, one pint flour, two eggs, one tablespoonful melted butter. Bake in hot iron gem pans.

JENNIE S. HARRISON.

Buttermilk Gems.

One quart of buttermilk, one heaping teaspoonful of saleratus, half a cup of sugar, five cups of sifted flour. Put in hot gem pans. Bake fifteen minutes in a quick oven.

MISS C. WELLINGTON.

Graham Gems.

One egg well beaten, two cups milk, one cup graham, one cup white flour, one spoonful sugar, a little salt, and a hot gem pan. MRS. WILLIAM C. COLLAR.

Graham Gems.

Two cups of graham, and one of white flour, one egg, a pint of new milk, a tablespoonful of sugar, a piece of butter as large as half an egg, melted, two teaspoonfuls baking powder, and salt to the taste. Beat well together half an hour before baking. Heat the gem pans very hot, butter them well, drop in the dough and bake in a hot oven fifteen or twenty minutes. MARY A. LIVERMORE.

Graham Gems.

Mix two parts Franklin graham, and one part fine flour, enough to fill your gem pan. Put in Royal Baking Powder, according to quantity (directions on each box), and a little salt. Beat up one egg, light, add milk enough to make a stiff batter. Butter the gem pan, let it stand on the stove until hot. Fill, and put into a hot oven. Will bake in fifteen minutes. MISS L. A. HATCH.

Graham Gems.

Two cups of milk or one cup milk and one cup water, a little salt, one tablespoonful of butter, a mixing spoon of molasses, and the measure of Horsford's Baking Powder; graham meal to make a soft batter. Bake in gem pans.

LOUISA G. ALDRICH.

Graham Gems

Take three cups of entire wheat flour or graham made from white wheat, two cups cold water, half cup of milk. Omit salt. Heat gem pans very hot on the top of the stove, fill them even full with the batter, place on the grate of a very hot oven. The "acorn" gem pans are essential. These are small, round, deep, iron pans. Notice, three things are necessary for good gems: The *best white wheat* flour, *very hot* pans and oven, and the "acorn" gem pans. No beating is required. These conditions observed, the gems will be as light as sponge cake. They can be eaten warm or cold, but are best heated over in a quick oven. They make excellent toast and puddings.

ALICE B. STOCKHAM, M. D.

Jennie's Sally Lunn Gems.

One egg, two tablespoonfuls of sugar, two of melted butter, one cup of milk, two cups of flour, one teaspoonful of cream of tartar, and one-half teaspoonful of saleratus. Bake in gem pans fifteen minutes. MRS. AUGUSTA RICH.

Oatmeal Gems.

Two and one-half large spoonfuls of oatmeal, one third cup of molasses, one pint of boiling water, mix, and stand over night. In the morning add one teaspoonful of cream of tartar, one small teaspoonful of saleratus, salt, and one-third quart of flour, or more, if needed. MISS M. A. HILL.

Oatmeal Gems.

To one cup of oatmeal add one quart of boiling water and two teaspoonfuls salt, simmer slowly about an hour; pour into a dish containing one tablespoonful of butter and two of sugar. Stir in flour while hot — a little at a time — making a stiff batter; when cold add a quarter of a compressed yeast cake dissolved in as little water as possible, cover and stand in a warm place to rise over night. Bake twenty minutes in hot gem pan. MISS C. WELLINGTON.

Corn Meal Griddle Cakes.

One cup of boiling milk poured on one cup of corn meal in which has been mixed one saltspoonful of salt, set this aside to cool; when thoroughly cooled add one cup of flour, one egg, a little sugar, one and one-half teaspoonfuls of baking powder. Beat the mixture *thoroughly* and set aside for ten minutes before frying; beat up the mixture well after each griddleful has been baked. M. F. D.

Indian-Meal Griddle Cakes.

Two cups of meal, one cup of flour, two-thirds cup of sugar, two tablespoonfuls of home-made yeast, wet with warm milk or water, in the evening. In the morning add three eggs and a little soda. MRS. SARAH S. SWAIN.

Snow Griddle Cakes.

Take six tablespoonfuls flour, add a little salt, and six tablespoonfuls of light, fresh-fallen snow. Stir the flour and snow well together, adding a pint of sweet milk. Bake the batter in small cakes on a griddle, using only a very little nice butter. They may be eaten with butter and sugar, and are very delicate. EDNAH D. CHENEY.

Squash Griddle Cakes.

One egg, one pint milk, one cup and a-half of boiled and strained squash, one-half teaspoonful soda, a little salt. Flour enough to make a batter. HULDA B. LOUD.

Pone.

Pour upon a half cup of sifted, salted meal, one pint boiling milk, and stir well. When partially cool beat in one egg, and pour into hot buttered pan. Bake half an hour in a quick oven.

Southern Pone.

Moisten two tablespoonfuls of sifted, salted Indian meal, with boiling milk. When cool, beat in one egg and a little cold milk. Drop the batter from a spoon upon the hot pan. Indian-meal as a rule should be mixed with boiling milk or water to allow it to swell, be well beaten, and free from all soda or baking powder. LUCY GODDARD.

Mother's Buns.

One cup butter, two cups sugar, one cup currants, one pint boiling water, one cup of yeast, or a yeast cake dissolved in a cup of water. Pour the boiling water on the sugar and butter, add the currants. When cool enough, stir in the yeast with flour enough to make it pretty thick (not so thick as bread). Set it it in a warm place, and let it rise about twelve hours. When well risen, make the quantity of buns desired, by moulding lightly in the hand, without adding flour. Let the buns rise well, and bake them in a moderate oven. When brown and done, brush the top with a syrup of milk and sugar (one-fourth of a cup of sugar to one-half a cup of milk). Do not not return to the oven. Keep the rest of the dough in a cool place. It will keep good for several days, and fresh buns can be baked every day.

HARRIET H. ROBINSON.

Muffins.

One pint milk, one cup sugar, five cups flour, one teaspoonful soda, two teaspoonfuls cream of tartar, two eggs, butter size of an egg. Bake quickly.

MRS. DR. FLAVEL S. THOMAS, M. S.

Muffins for Breakfast.

One quart of flour, one tablespoonful of yeast, mix with cold water to a stiff batter. Set in a warm place over night. Next morning add two eggs, beat it well with a spoon. Bake in rings in a hot oven.

MRS. M. ANGELO FOSTER.

Breakfast Muffins.

Two cups flour, into which stir two teaspoonfuls baking powder, one cup milk, one tablespoonful butter, one tablespoonful sugar, and two eggs. Bake in muffin rings or gem pans.

MRS. MARTHA J. WAITE.

Graham Muffins.

Take one pint of new milk, one pint graham flour, or entire wheat flour. Stir together, and add one beaten egg.

Can be baked in any kind of gem pans or muffin rings. Salt must not be used with any bread that is made light with egg. ALICE B. STOCKHAM, M. D.

Graham Muffins.

One egg, one-quarter cup molasses, two and one-quarter cups sour milk, one-half an even teaspoonful of soda, two even cups of graham, and one even cup of flour. Add the egg, well beaten, to the molasses and sour milk, reserving a tablespoonful of the milk with which to dissolve the soda. Add the soda when dissolved, and afterward the graham and the flour. If you have no sour milk, add a teaspoonful of vinegar to the milk to sour it. Bake in muffin rings or a gem pan. They are delicious. MRS. O. A. CHENEY.

Oatmeal Muffins.

In a quart of well-cooked oatmeal pudding, warm, stir a small piece of butter and half a cup of sugar. Put in half a cake of yeast, and mix stiff with flour. Raise over night and bake in gem pans. ANNE B. ROGERS.

Raised Muffins.

One pint of milk, one quart of flour, six eggs (five will do), piece of butter size of half an egg, one teacup of sugar, two tablespoonfuls of yeast, or one-sixth of a compressed yeast cake. Put the butter into the milk, warm it until the butter melts, then pour it gradually to the flour, mixing smoothly. Add the other ingredients, and set it to rise over night. In the morning add a little salt. Bake in muffin rings about half an hour. MRS. ELIZA B. BURGESS.

Rye Muffins.

One and one-half cups flour, one and one-half cups rye meal, one tablespoonful sugar, two teaspoonfuls cream of tartar, one teaspoonful soda, two cups milk, two eggs, and a little salt. Bake in gem pans.

MRS. WILLIAM C. COLLAR.

Pancakes.

One cup of buttermilk or sour milk, two cups of sugar, two eggs, half spoonful saleratus, teaspoonful of salt, pinch of cinnamon, five scant cups of sifted flour; drop by tablespoonfuls into boiling fat, cook until nicely browned, about ten minutes. MISS C. WELLINGTON.

Rye Pancakes.

One cup of milk, one egg, one teaspoonful of soda, a little salt, and a small half cup of molasses. Beat all together, then stir in two-thirds rye-meal and one-third flour, sufficient to make a batter that will drop from the spoon readily into the hot fat. Fry as you would doughnuts.

LOUISE G. ALDRICH.

Pop Overs.

Two cups milk, two cups flour, two eggs, a little salt. Bake quickly in gem pans. MRS. M. A. EVERETT.

Water Puffs.

Two eggs beaten into one pint cold water, add one-half teaspoonful salt and one pint sifted flour. Bake in generously buttered gem pan, half an hour. (Excellent with coffee for breakfast.) MISS H. B. HICKS.

Corn-Meal Rolls.

Two cups meal, one of flour, three cups of milk, a tablespoonful of sugar, one teaspoonful of soda, one of salt, two of cream of tartar, a well beaten egg. Bake in roll-pans.

JANE L. PATTERSON.

Parker House Rolls.

Boil one pint of milk, add a piece of butter the size of an egg, and let it cool. Add half a yeast cake dissolved in warm water. Mix in flour, the same as for bread, then knead twenty minutes. Mix early in the morning for tea. Roll out two hours before baking. Cut out with a biscuit cutter, put a little butter on, and fold over half the roll. Bake quickly.

ZILPHA H. SPOONER.

Rye Rolls.

One-half cup of Indian meal, one cup of rye-meal, two of flour, one-half cup of sugar. Make a little stiffer than griddle cakes, raise with yeast. Scald the Indian meal, and let it swell before putting in the other ingredients. Add about half a teaspoonful of dissolved soda in the morning, and bake in roll-pans.

ANNE B. ROGERS.

Sally Lunn.

One quart of flour. Stir into it two teaspoons even full of cream of tartar, and one teaspoonful of soda or saleratus. Stir in one egg, one tablespoonful of sugar, one cup of milk, a little salt.

MRS. EMILY A. FIFIELD.

Squash Cakes.

One pint of squash cooked and strained; while warm stir in two tablespoonfuls of sugar, one tablespoonful of butter, one teaspoonful of salt, one-half teaspoonful of soda made into a smooth powder, one-half cup of good yeast. Flour to roll soft. Mix after dinner for breakfast, or at breakfast-time for tea, as they rise slowly. These are delicious when just right, and worth trying till right as to the quantity of flour. Roll and make into biscuits. A well tried receipt for more than fifty years.

MRS. SARAH R. MAY.

Corn Muffins.

One-half cup fine white corn-meal, one cup white flour, one and one-half teaspoonfuls Royal Baking Powder, one tablespoonful sugar, butter size of an egg, one-half pint sweet milk. Bake in roll-pan, in a brisk oven.

MRS. MARY CROSS HARRIS.

Strawberry Short Cake.

One-half cup sugar, one cup butter, one pint milk, one teaspoonful soda, one and one-half pints flour, two teaspoonfuls cream of tartar, salt. Bake and split, insert berries previously mashed and sugared, and put a few large berries on top. Eat warm. MISS L. F. S. BARNARD.

Rhubarb Toast.

Take one pint water, half a cup of sugar; when boiling put in two pounds rhubarb cut in small pieces. Stew until done, when cold pour over a platter of hot toasted graham bread, having a little butter upon it. This is an excellent breakfast dish, and as the toast absorbs the peculiar rhubarb flavor, can be eaten by those who usually dislike it. Gooseberries and tart apples can be prepared in the same way. *Note.*—Never use white bread for toast, when bread of the unbolted or entire wheat flour can be had. The latter never becomes doughy, and is much better flavored, besides being more nutritious. ALICE B. STOCKHAM, M. D.

Golden Toast.

Soak slices of dry bread in milk till thoroughly wet, but not too soft, dip into beaten eggs, and fry in butter.

LILLIE DEVEREUX BLAKE.

Waffles.

One pint sweet milk, one and a half cupfuls of sour cream, three eggs — whites beaten separately, two teaspoonfuls of baking powder, flour to make a thick batter.

EMILY S. BOUTON.

Waffles.

One pint of scalding milk poured over one teaspoonful of sugar, one tablespoonful of butter, and a little salt; when lukewarm add two tablespoonfuls of yeast, and flour enough to make the batter of the consistency of thick cream; set to rise over night, and just before baking add two well-beaten eggs. If any flour is added after the batter is raised, they are sure to be tough.

EGGS AND OMELETS.

Breakfast Dish.

Cut smoothly from a wheaten loaf
  Ten slices, good and true,
And brown them nicely, o'er the coals,
  As you for toast would do.

Prepare a pint of thickened milk,
  Some cod-fish shredded small;
And have on hand six hard-boiled eggs,
  Just right to slice withal.

Moisten two pieces of the bread,
  And lay them in a dish,
Upon them slice a hard-boiled egg,
  Then scatter o'er with fish.

And for a seasoning you will need
  Of pepper just one shake,
Then spread above the milky juice,
  And this one layer make.

And thus, five times, bread, fish and egg,
  Or bread and egg and fish,
Then place one egg upon the top,
  To crown this breakfast dish.

ELIZABETH W. STANTON.

Waterlily Eggs.

Boil two eggs twenty minutes. Separate whites from yolks. Put on a plate one teaspoonful of flour, a piece of butter the size of a hickory nut, and pepper and salt to taste. On this plate cut up the whites of the eggs into small cubes the size of dice, mixing with flour, salt, etc. Have four tablespoonfuls of milk boiling in a sauce-pan; put the whites in, and let them cook slowly while you make two slices of toast. Spread whites (when flour is thoroughly cooked) over toast. Break the yolks up slightly and salt them, and force through fine strainer over the whites on top of the toast. Holes in strainer should not be larger than pinheads. Serve hot, at once. A very pretty dish, and convenient in case of unexpected company, as bread and eggs are almost always in the house.

ALICE STONE BLACKWELL.

Omelet.

Beat the yolks of two eggs, and stir in five tablespoonfuls of milk and a pinch of salt, then beat the whites of two eggs very light, and stir all together. Heat the omelet pan hot, and then put in a little butter, and when melted turn in the beaten mixture; set on the fire and cook until a light brown, then fold the omelet and serve on a hot dish.

LOUISA G. ALDRICH.

Omelet.

Wet one tablespoonful of flour with milk and stir till a smooth paste is produced, add the yolks of three eggs, beat well; to this add one tablespoonful of melted butter and one teacupful of milk; beat the whites to a stiff froth and stir well through the mixture. Butter a good-sized frying-pan, and have it hot to receive the omelet; cook five to seven minutes. To take it up, use a large knife and turn one half over the other, and turn out the whole on a platter.

MARY GAY CAPEN.

Egg Omelet.

Three eggs, two tablespoonfuls cream, and a pinch of salt and pepper. Butter omelet-pan first with beef suet, to prevent sticking, then put in a piece of butter the size of a nutmeg; put in the omelet. Remove it from the pan before it browns. JENNIE S. HARRISON.

Egg Omelet.

Three eggs—whites and yolks beaten separately, half cup milk heated, one tablespoonful flour; thicken and add butter the size of a walnut; when cool add the yolks, with a little pepper and salt; add the whites beaten stiff, and cook in buttered spider until done, about two minutes.

MRS. H. ANDREWS.

Plain Egg Omelet.

Six eggs, two tablespoonfuls flour, one teacupful of milk, one teaspoonful of salt; fry in butter, a piece as big as an egg. Beat the yolks separately, stir in the flour, then add the milk, and the whites beaten to a smooth froth. To fry in butter gives it more of a relish than in lard. The above will make two omelets fried in a common-sized spider. It is a good plan to set it in the oven before turning out on to a platter. KATHARINE STARBUCK.

SOUPS.

Potato Soup.

Mash boiled potatoes perfectly smooth, with butter and salt, and pepper if wished; split stalks of celery and cut in small pieces and boil till tender; add to the potato a little of the water in which the celery was boiled, and mix it well with the potato; mix boiling milk with the potato, and strain it all; then add the celery, and boil it up once.

LUCY GODDARD.

Soups.

Put a piece of good, juicy beef, say four or five pounds, into a middling-sized pot, with one cupful of oatmeal, one or two onions, pepper and salt, boil until the beef is tender. It is a good plan to let the soup boil hard for a short time, then set it back to simmer until done. Strain.

Rice, Vermicelli, Sago, raw Potato (grated) and Farina soups can all be made in the same way. Tomatoes or celery give a pleasant flavor, and help to make variety in soups.

For Vegetable soup, instead of oatmeal, etc., cut up cabbage, potatoes, carrots, white turnips (less of the carrots and turnips). MRS. L. B. URBINO.

Bean Soup.

Take one pint white navy beans, put over the fire in cold water, with a small pinch of saleratus; when the water comes to a boil drain off, fill up with boiling water, add a two-inch cube of fat pickled pork, a teaspoonful of salt, and a little pepper; after boiling about two hours, strain off the soup, mash the beans and stir them in, cut meat in small squares, and serve. Add hot water, if needed, in cooking.

MRS. JESSIE F. A. BANKS.

Beef Stew.

Three pounds beef (round) cut in inch pieces, four or five onions peeled and sliced. Put in a layer of meat, then a layer of onions. Dredge well with salt, pepper and flour. Repeat until all the meat and onions are used. Add two quarts boiling water, and simmer three hours. Then add one quart of potatoes peeled and sliced, and three tablespoonfuls flour mixed with one cup of cold water. Simmer thirty minutes longer.

JOSEPHINE P. HOLLAND.

Chicken Soup.

Cut the chicken as for a stew, after thoroughly washing, place in an earthen dish, cover with ice-cold water, slightly salt, and let it stand two hours before putting it into the soup-kettle. Place in a porcelain-lined soup-kettle, cover with two quarts of water and let it come to a boil slowly; keep it at the boiling point until the meat is so tender that it can easily be removed from the bones; take the meat from the kettle, season the liquid with salt to taste, and with two teaspoonfuls of butter, and place it (closely covered) where it can simmer without coming to a boil. Then remove the meat from the bones, and chop the white and the dark parts of it separately; half of the white meat from one fowl and an equal amount of the dark meat will be sufficient; mix these meats, and with them mix half of the quantity (measuring with a tablespoon) of very finely rolled cracker crumbs, (graham crackers preferred); season with two teaspoonfuls of butter, a "pinch" of salt and a fourth of a teaspoonful of grated nutmeg; bind the mixture together with the white of one egg well beaten, and mould firmly into hard balls about three-fourths of an inch in diameter. Place the balls in the liquid, and let it come to a boil; lift with a skimmer the balls from the kettle into the tureen, pour the liquid over them, grate a little nutmeg over the top, and sprinkle upon it a little parsley broken into small bits. This soup gives out an appetizing odor, and to the average palate is not disappointing.

MAY WRIGHT SEWALL.

Excellent Veal Soup.

Put a knuckle of veal into four quarts of cold water, add three teaspoons of salt, and boil for an hour. Have ready three large onions, three turnips, two large carrots, three or four tomatoes, a head of celery and a bunch of sweet herbs. The onions, carrots and turnips must be chopped very fine. Put all into the soup, and let boil an hour longer. Then, take out the meat, and strain the soup. Cream half a cup of butter with a large spoonful of flour; stir into the soup, return to the fire, and let it boil a few minutes. A cup of good cream, scalded, and added at the last, is a great improvement. CATHERINE H. BIRNEY.

Mock-Turtle Soup.

Put into a pan a knuckle of veal, two calves' feet, two onions, a few cloves, peppers, berries of allspice, mace and sweet herbs; cover them with water; cover the pan with thick paper, and set it in the oven for three hours. When cold take off the fat, cut the feet into bits an inch and a half square, remove the bones and coarse parts, then put the rest on to warm with a large spoonful of catsup, and the jelly of the meat. Add more seasoning if liked, and serve with hard boiled eggs, force meat balls and a little lemon juice. This is easy to make, and the dish is excellent.

MISS E. B. PLYMPTON.

Pea Soup.

Soak a quart of split peas over night in cold water. Put them on to boil early, and stir them frequently. Season the soup with the liquor in which corned beef has been boiled, adding it after the peas are well cooked, in just sufficient quantity to make the soup agreeably salt. Strain smooth, and serve with cubes of brown toast.

EDNAH D. CHENEY.

Potato Chowder.

Peel and slice a dozen potatoes, and put them into cold water. Put a generous tablespoonful of butter and a quart of hot water into a sauce-pan, add a scant teaspoonful of salt, and when boiling hot add the potatoes, and cook slowly three-quarters of an hour. Add a pint of milk and let it just come to a boil, add more salt if necessary, a little pepper if liked, and serve immediately.

MRS. E. E. KELSEY.

Potato Soup.

Take one quart milk, into which put a good sized onion, let it boil till tender, then remove it. Pare three medium sized white potatoes, and boil them separately. Strain or sift them through a cullender into the hot milk, and let it come to a boil. Have ready your soup dish, into which place a tablespoonful of butter, and salt to taste. It is preferable to use a double boiler. MRS. MARTHA J. WAITE.

Rebel Soup.

Heat one quart of milk to the boiling point, add one cracker rolled fine; to one cup of tomatoes add one-fourth teaspoonful soda, stir, and while foaming add it to the boiling milk; put butter, salt, and pepper in the dish, and pour the soup on them. MRS. MARY F. CURTISS.

Red Bean Soup.

Take one quart red Italian beans, soak and rinse twice in cold water, and boil in third water (about two quarts) four or five hours; brown one onion, some celery, and carrot in butter, and add one hour before the soup is done; strain as usual, and add one hard-boiled egg cut up, and half a lemon. EDNAH D. CHENEY.

Tomato Soup.

Put one pint of canned or fresh tomatoes and one quart of water in a Granite stew-pan; when boiling, thicken with three teaspoons of graham flour mixed with cold water; add one quart milk, and stir until it boils — this prevents curdling; season to taste. ALICE B. STOCKHAM, M. D.

Tomato Soup.

One can tomatoes, half a can of water, a quarter of an onion; boil twenty minutes, strain, and add salt, pepper, a small piece of butter, and half a cup of milk; then return to the stove and thicken with about a teaspoonful of corn starch. A little red pepper or a clove or two improves it.

HULDA B. LOUD.

FISH, OYSTERS AND CLAMS.

To Make Mauve, or Soft Clam Soup.

Pour boiling water over the clams, and, after opening, separate the pulp and cartilage, cut the latter into small pieces and boil twenty minutes, add the pulps, season with salt, cayenne and black pepper, thicken with flour rubbed in butter, and a pint of cream or rich milk; serve as soon as it boils. Be careful to preserve the liquor when opening the clams, and use it for the soup.

ANNA ELLA CARROLL.

Boston Fish Chowder.

Three and a half pounds of haddock—head and all, three thick slices salt pork, one dozen Boston crackers, eight large potatoes, one onion, one pint milk, one tablespoonful salt, one-half teaspoonful pepper, flour; order the fish skinned and cut up in about eight pieces, exclusive of the head; fry the pork slowly in rashers, pare and slice the potatoes, chop the onion fine. Use a good-sized kettle that can be tightly covered, and place the ingredients in the following order: a little of the pork fat, a layer of fish, a layer of potatoes, with salt, pepper and flour sprinkled over the top; repeat this till all the fish and potatoes are used up, then pour boiling water enough to cover over the whole; cook over a slow fire till the potatoes are done; scald the milk and add; split the crackers, put them in the tureen and saturate them with boiling water. Take the chowder up carefully with a skimmer, removing the head and the loose bones, and put it with the crackers; try the juice to see if it is seasoned to suit; if not thick enough, add a little flour; strain and pour over the whole; serve the rashers separate.

HARRIET H. ROBINSON.

To Broil White Fish.

Among the fish of our great lakes, the white fish hold the highest rank with epicures. Those of Superior, where the water is never warmer than 40°, are the best, because of the hardness and firmness of their flesh. Upon the shores of this great inland sea I learned—a Chippewa Indian's receipt—that, to have this fish in perfection, it must be covered while broiling. Having prepared your fish, salting it somewhat, place it on the gridiron over the fire (of course flesh side down), and cover it with the dripping-pan; when cooked upon this side and turned for the slight finish requisite, with a round-ended knife (to avoid breaking the flesh) press down a bit of nice butter here and there over the fish. It is then ready for the table, and should be eaten at once. A white fish thus cooked tastes quite unlike one prepared by ordinary methods.

MATILDA JOSLYN GAGE.

Fish Balls.

Pare six large potatoes, and boil with one pint salt fish pulled into small pieces, then mash together; beat separately the yolks and whites of three eggs (or two will do), stir together and add to the potato; mould in balls in your mixing spoon, and fry in hot fat. MISS H. B. HICKS.

Lobster Soup.

Take one large or two small lobsters, pick all the meat from the shell and chop it very fine, season it with salt and cayenne pepper to taste; pour into a pail or pitcher placed in a kettle of boiling water, one quart of new milk; before the milk comes to a boil make a batter of one-quarter pound of butter and a tablespoonful of flour, and stir it into the milk, then add the lobster, and boil ten minutes.

MARY C. AMES.

Escallopped Oysters.

Soak cracker-crumbs in milk. In the bottom of a pudding-dish put a layer of the cracker-crumbs, on this a layer of oysters, seasoning with pepper, salt, a little lemon-juice or vinegar, plenty of butter, adding some of the oyster liquor; on this put another layer of soaked crumbs, then one of oysters well seasoned, continuing these layers till the pan is full, a layer of crumbs being on top. Brown well, and serve hot. MISS L. F. S. BARNARD.

Macaroni and Oysters.

Take boiled macaroni and put a layer in a deep dish, above this put a layer of good-sized oysters dried with a soft towel; season these two layers with butter, pepper and a very little salt; add another layer of macaroni, season with butter and salt; a layer of oysters, season with butter, and *pepper*, and salt; the top layer of macaroni, with butter and salt. Set in the oven long enough to cook oysters and brown the macaroni.

MRS. SARAH R. BOWDITCH.

Salmon Hash.

Mash until light eight good-sized potatoes, season thoroughly, stir into the potato one-half can of salmon picked fine; heap on a platter, smooth, and mark with a fork, and set in the oven to brown. Salt salmon may be used instead of canned salmon. ELLA C. ELDER.

Stuffed Lobster.

Chop very fine, then add to a lobster weighing three pounds and a half, a piece of butter the size of two eggs, salt, two handfuls of bread crumbs, a little milk, nutmeg, black pepper, and a little cayenne; then put it on the fire for a few minutes until heated through, then put in shells. cover with crumbs, add a little butter, and brown in the oven. REBECCA HOWLAND.

To Cook Terrapin.

Decidedly the terrapin has to be killed before cooking, and the killing is often no easy matter. The head must be cut off, and, as the sight is peculiarly acute, the cook must exercise great ingenuity in concealing the deadly weapon. I have known half an hour to be consumed in the effort. When accomplished, the terrapin is put in cold water and boiled until the feet can be easily pulled off. Remove it from the water, take off the bottom shell, separate the four quarters, be careful to take the gall from the liver, then utilize every part but the sand bag and the intestines; season with cayenne, salt, and a little black pepper; put in a small quantity of the water in which it was boiled, add three or four ounces of butter and a pint of good cream. Stew for ten minutes. ANNA ELLA CARROLL.

MEATS.

COOKING MEATS, AND LENGTH OF TIME REQUIRED.

Boiling Meats.

Meat for soups, or whenever it is to be served in the liquor in which it is boiled, should be put first into cold water and gradually brought to the boiling point, and allowed to simmer gently till done — never boil *hard.*

Meat to be served separately, without the liquor, should be put at once into boiling water, to harden the surface and prevent the juices from being drawn out, brought quickly to the boiling point, and allowed to simmer gently till sufficiently cooked. Nothing is gained by rapid boiling, as the meat cooks no quicker, and much of the flavor is lost by being carried off in the steam. The time required for corned beef is four hours; mutton and lamb, three hours; veal, two and one-half hours; chicken, two hours; fowls, three hours; corned pork, three hours.

Roasting Meats.

Beef, from one to two hours, according to the size; mutton and lamb, three hours; veal, two hours; chickens, two and one-half hours; turkey, three hours; goose and ducks, parboil one hour before roasting, then roast one and one-half hours; pork, three hours. For roast meats have a quick oven.

Irish Stew, or Pot-pie.

Two pounds of veal cut in six pieces, or more, put it on in two quarts of cold water and let it cook slowly until nearly done; add a dozen or more potatoes, whole; boil until they are half done, add salt and two onions, sliced.

For the paste, or upper crust, take one large pint of flour, one heaping teaspoonful of baking powder, a small piece of butter (put in as for biscuits), one egg, mix them together with milk or water, roll out thick as for biscuit, the size of the top of your kettle, then cut it in four pieces and put it on top of your stew; keep the cover on, and let it boil slowly for fifteen minutes. Three minntes before serving, and while boiling, mix one heaping teaspoonful of flour with cold water, put this in your stew to thicken the gravy. Keep at least a quart of water in your kettle all the time, and be careful that it does not burn. It may be well to put a plate in the bottom of the kettle, to keep the stew from burning. If the water gets low, add boiling water at any time. Chicken or beef, instead of veal, is very good to be used. Serve this on a large platter; do not divide it in different dishes. CORA SCOTT POND.

Veal Loaf, or Frigadelle.

Three pounds of nice lean veal, one-quarter pound of fat salt pork, chop them very fine, add a teacup of pounded cracker and one egg thoroughly beaten, a heaping teaspoonful of salt, a teaspoonful of black pepper, one ounce marjoram or sage; all these to be thoroughly mixed with the hand, and put into a bread-pan and baked one hour and a quarter. Very nice cold, for lunch or tea. MARY C. AMES.

To Broil Beef Steak.

Have the steak cut three-quarters inch thick, remove all the fat portions (these may be *separately cooked* if desired).

Give the meat a sprinkling of salt, then of pepper, and last, a dredging of flour, before putting into the broiler. Cook with intense heat and quickly. The steak should be turned once.

MRS. MARY S. TARBELL.

Roast Beef with Yorkshire Pudding.

Set a piece of beef to roast upon a grate laid across the dripping-pan (a large wire broiler answers very well). Three-quarters of an hour before it is done, mix the pudding and pour into the pan under the meat, letting the dripping from the meat fall upon the pudding. When the pudding is done, cut into squares and serve with the meat. If there is much fat in the pan before putting in the batter, pour it off into a cup and use it to baste the meat with.

Yorkshire Pudding.

One pint of milk, four eggs, whites and yolks beaten separately, two cups of flour, one teaspoonful of salt. Be careful in mixing not to get the batter too stiff.

MRS. M. M. WOOLFORD.

Boiled Tongue.

Purchase a salted tongue and boil six hours. Pare while hot, and put in a cool place. Slice thin for the table. When cooked in this way it is a real delicacy.

MRS. SARAH M. PERKINS.

Chicken Croquettes.

The meat of one chicken (no skin) chopped fine, one-third the quantity of fine bread brumbs, one-third cup butter, one tablespoonful of salt, one-half teaspoonful of pepper — or, better still, "season to taste," — three medium-sized boiled potatoes chopped very fine. Mix meat, crumbs and potatoes, then season and add the melted butter, and mix it well in; then add milk enough to make it quite moist, but still so that the croquettes will stay in form; roll in beaten egg and then in cracker-crumbs and fry in hot lard, and if you wish complete success use always a wire frying-basket.

DR. LEILA G. BEDELL.

Chicken Croquettes.

One good sized fowl, one-quarter pound butter, one-half pint sweet cream, three tablespoonfuls flour, one-quarter pint chicken stock, a little ground nutmeg, salt and pepper to taste, and the juice of half a lemon. Boil the chicken and let it cool, remove the meat from the bones and cut in small pieces. Melt the butter in a saucepan, stir in the flour, cream, and a little of the stock; after stirring for a few moments take it off the fire and add the chicken and seasoning. Spread on a platter to cool, and when quite cold shape in the form of pears, dip them in egg and cracker-crumbs and fry in boiling-hot lard till they are of a nice brown. This recipe can be used for any other kind of meat.

MR. C. M. RANSOM.

To Broil Chickens.

Cut the chickens open on the back, place them on the meat-board and pound until they will lie flat on the gridiron, then broil them over hot coals for twenty minutes or more, until they are a nice brown; turn them frequently, and do not burn them. Put the liver, heart and gizzard in a stewpan, add a pint of water for each chicken and boil until they are tender and can be chopped very fine. Then to this add butter, pepper, salt, and a little flour for thickening, with a cup of sweet cream if you have it. When the chickens are done, dip them while hot in this gravy, put them back on the gridiron over the coals for a minute, taking care that they do not burn; then place all in the gravy, allowing it to boil up once, and send to the table hot.

MARY A. LIVERMORE.

To Fry Spring Chicken and Make Gravy as Mother Did It.

Cut the chicken into pieces convenient for serving. Have ready a frying-pan half full of boiling lard or butter. Roll the pieces, first slightly salting them, in fine flour or corn meal. Fry quickly till thoroughly done; dish on to a large platter and pour the surplus "fryings" into a bowl for future use, leaving less than a gill in the bottom of the frying-pan. Into this stir rapidly a heaping tablespoonful of flour, then add a pint of fresh milk, stirring constantly till it boils and thickens. Salt and pepper to taste. Pour the gravy over the chicken in the platter, or, if preferred, in a separate gravy dish. Any kind of young fowl is delicious cooked in this way, and no child forgets the delights of the side dish of gravy that accompanies it. ABIGAIL SCOTT DUNIWAY.

Chicken Jelly.

Prepare a chicken as if for roasting; lay aside the liver, gizzard, heart and neck; singe it; wash it in cold water lightly, but do not wipe it. Cut the legs and wings off, and cut the body in two, lengthwise. Put all the parts into a common china ginger-jar with three cloves, three whole peppers and a blade of mace, all just once cracked. Set the jar, covered with the china cover, into a kettle of water and let the water *boil gently* about it for eight hours; take the kettle off and raise the jar, turn out your chicken on a dish and pick out all the skin, bones and sinews; have a bowl with cold water in it ready. When the picking is done salt the remainder to taste, put it all into the bowl and cover with a plate. It will turn out next day in shape and is nice eaten cold with bread, or made into soup with a little boiling water as needed. MRS. SARAH R. BOWDITCH.

Deviled Kidneys.

Take two beef kidneys, cut them into small squares, excluding the fat; pour over them boiling water; let them stand a few moments to extract the blood, then put them into your stew-pan, covering them completely with water. Boil one hour, then add one large onion, chopped fine, a tablespoonful of sage, a dessert spoonful of each summer savory and sweet marjoram, a pinch of clove. Cook gently for two or three hours, and when ready to serve cut slices of lemon into your dish, squeezing out some of the juice in the dish; pepper and salt to taste, pour the stew over the sliced lemon, add butter the size of an egg.

MRS. M. M. WOOLFORD.

Fricandelles.

Chop cold meat very fine, add teacup of scalded milk in which a teacup of bread crumbs has been rubbed smooth, half cup of butter, juice of half a lemon, salt and pepper. Make into balls, roll in yolks of eggs, brown in butter. Remove balls, brown tablespoonful of flour in the butter, add slowly a pint or more of beef stock, boil two minutes, replace balls and cook slowly for an hour. Serve with toast and lemon. Beef, chicken, etc., may be used. If veal is used, add half cupful chopped ham.

MRS. JESSIE F. A. BANKS.

Scalloped Mutton.

Put into a deep dish a layer of cracker crumbs, a layer of sliced mutton, a layer of stewed tomato, a layer of cracker crumbs, a layer of macaroni, a layer of mutton, and finish with a layer of cracker crumbs. Season with pepper, salt, and a little butter, and moisten well with gravy or stock. Beat an egg and pour over the top. Bake twenty minutes.

ELLA C. ELDER.

Newburyport Housekeepers' Way to Glorify Cold Mutton.

Take a well buttered pudding-dish and put into it a thick layer of chopped cold mutton salted to taste, then a thinner one of canned tomato, salted, and so on until the top layer, which should be tomato with cracker crumbs thickly strewed on it and bits of butter dotted here and there. Cook in the oven until thoroughly heated through and well browned on top. It is well to cover at first, then uncover just long enough to brown. MRS. ELLEN W. E. PARTON.

Sausage Meat.

For twenty pounds meat: half pound salt, one small cup pepper, one large cup sage, one small teaspoonful saltpetre.

MISS E. M.

Baked Sausages.

Put them in a baking tin in the oven, turning frequently as when frying. Brown them well. They are less greasy than if fried, and are more delicate in every way.

MRS. M. F. WALLING.

New England Sausages.

To thirty pounds of sausage meat add ten ounces of salt, three ounces of sage, one and one-half ounces of pepper, two tablespoonfuls of alspice and one cup of sugar. Let the measures be exact. . Do not allow the meat to freeze, but as soon as possible cut it in pieces about an inch square; scatter the seasoning over and mix it in with the hands in a large pan, wooden bowl, or, if a very large quantity, on a cooking-table. Chop or grind the meat very fine. Sausage meat is sometimes packed in jars and melted lard is poured over the top to exclude the air, but the better way is to make some long bags of strong cotton cloth of such a size that when filled they will be as large round as a teacup. Dip the bags in strong salt and water, and dry thoroughly before filling. They should be crowded full and each end tied up. When the meat is to be used, open one end of the bag, rip the seam a little way, turning the cloth back, cut off slices rather more that an inch thick, dip in flour and fry.

MRS. ALLIE E. WHITTAKER.

Scalloped Meat.

Put in a deep dish alternate layers of mashed or finely chopped potato and chopped meat. Season each layer and moisten with gravy or soup stock. Use potato for the top layer, pressing it through a sieve. Bake twenty minutes. Any cold meats may be used in this way.

ELLA C. ELDER.

Veal Loaf.

Three pounds raw veal, one-quarter pound of salt pork chopped fine; mix with two eggs, one cupful of pounded crackers, two teaspoonfuls salt, two teaspoonfuls pepper, and a handful of fine-cut parsley or sweet marjoram; mix well, and press hard in a pudding-dish. Bake one and one-half hours. When cold, slice thin for tea.

MRS. D. W. FORBES.

Veal Loaf.

Three pounds finely-chopped veal, three eggs, eight water or butter crackers, two tablespoonfuls of cream, three teaspoonfuls of salt, one teaspoonful of powdered sage, one tablespoonful of black pepper, one lemon grated; mix well together and form into a loaf, using a little flour, put butter on top, baste often, and bake two hours.

MRS. THOS. S. LYON.

Veal Sausage.

Veal (without gristle) and a little fat pork chopped fine, one-half the quantity of bread-crumbs, pepper, salt, sage and thyme. To four pounds of veal, one spoonful of sage and thyme, one raw egg, and the bread; mix thoroughly, make into balls the size of a sausage, dip into crumbs, fry in hot suet twenty minutes, or till done. JULIA A. KELLOGG.

Veal Loaf.

Three pounds raw veal chopped very fine, one-half pound raw salt pork chopped fine. Add three well-beaten eggs, one teacup powdered cracker, one and one-half teaspoonfuls salt, four teaspoonfuls of mixed sage, savory, salt, and pepper. Then add four hard-boiled eggs by putting first a layer of the meat, then two eggs at equal distance apart, then another layer of the meat, then the two remaining eggs; then cover all with what meat remains. Bake two hours in a moderate oven, with the pan set in another filled with water. When thoroughly cold and sliced, it makes a palatable and pretty dish. LINDA S. BARNEY.

Curry of Mutton.

Two pounds of cold mutton, either boiled or roasted, juice of half a lemon, half pint of cream, three onions, four tablespoonfuls of butter, one tablespoonful curry powder; salt to taste. Pare and slice the onions, put the butter into a saucepan, when hot add the onions, fry a delicate brown, then add the meat, and cook until brown. Bring the half pint of cream to a boil, thicken with one tablespoonful of flour that has been rubbed smooth in a little cold milk. Add the curry powder, and salt to taste, pour over the meat and cook five minutes more (slowly), and serve hot with boiled rice. ALICE A. GEDDES.

VEGETABLES AND SIDE DISHES.

Boiling Vegetables — Length of Time Required.

Cabbage and beets, three hours; turnips and carrots, one and one-half hours; potatoes. squash and parsnips, about an hour.

Summer Vegetables.

Green peas, beets, turnips and greens, one hour; string beans, from three to four hours; squash and potatoes, three-fourths of an hour; green corn, one half hour.

Carrots.

Cut carrots into rather small pieces, boil until nearly soft, pour off the water and add salt, a bit of butter, and a little sugar. Dredge on flour and cook until done.

MRS. L. B. URBINO.

Red Cabbages.

Cut cabbage into eighths, boil in water with a little salt; when nearly done, pour off the water and put in a little vinegar and sugar; simmer until the cabbage is soft.

MRS. L. B. URBINO.

Macaroni a l'Italian.

Strain all the juice from a large can of tomatoes; add to the juice one tablespoonful Lucra oil, one-eighth pound grated cheese, one teaspoonful essence of celery, salt, and a mere dust of cayenne. Bring to a boil, and thicken to a pudding-like consistency with cornstarch blended in cold milk. Boil one-third package Italian macaroni in equal parts of salted milk and water until soft. Cut one pound tender and lean beef into dice-shaped bits, and fry in butter, with one boiled onion, adding salt and pepper. Serve in a hot dish, putting in first the tomato, then the macaroni, and lastly, the meat.

EVELYN GREENLEAF SUTHERLAND.

Scalloped Onions.

Put medium-sized onions into cold, salted water and boil one hour; drain off water; butter a baking-dish lightly, and line with stale bread or cracker crumbs. Put in a layer of onions, and sprinkle crumbs over all, adding pepper and salt to taste. Put a small piece of butter on top of each onion. Add a tablespoonful of water, and bake three-quarters of an hour till brown. Serve hot.

ALICE STONE BLACKWELL.

Fried Tomatoes.

Peel the tomatoes and cut them into two slices each. Have hot in the frying-pan a mixture of suet and butter in equal parts. Brown the tomatoes on both sides and take them out dry on a hot platter. Leave a small quantity of fat in the frying-pan; into this dredge a little flour and then add enough cream or rich milk to make a gravy to cover the tomatoes. Stir till well thickened, and pour over the tomatoes.

MARTHA B. PITMAN.

Noodles.

Take two eggs and sufficient flour to make a dough which can be rolled out. Sprinkle the board well with flour and roll out the dough as thin as possible; let it remain for an hour, or until somewhat dry. Now roll up and shave off with a knife, beginning at the end. The dough will then be in long strips which may be cut into shorter ones if desired. These strips are to be cooked in a very hot soup a few minutes before serving. MRS. ELIZABETH GROSHAN.

Home-made Macaroni.

Beat one egg well, add a pinch of salt, and half a coffee cup of cold water; then stir and mould in as much flour as the liquid will take up. The dough must be so stiff that it can with difficulty be rolled; roll out as thin as possible, scatter flour over it, and leave it an hour or two to dry, then roll up, cut in narrow strips, shake these out, and boil and prepare as other macaroni. If the paste is stiff enough and rolled thin enough, it will be found superior to the bought macaroni, and far cheaper. CATHERINE H. BIRNEY.

Boiled Potatoes.

Potatoes should never stand in water. They should be washed as quickly as possible and put into boiling water; they should not be entirely covered with water, and should cook very fast. As soon as a fork will pass through them, drain, cover closely and steam a few minutes. Serve as soon as done. MRS. D. W. GAGE.

Escalloped Potatoes.

Pare and slice thin; butter an earthen dish, put in a layer of potatoes, season with salt, pepper, butter, and a bit of onion chopped fine, sprinkle a little flour; in this way add layers of potatoes and seasoning until the dish is full; add a cupful of milk, and bake three-quarters of an hour.

MISS L. A. HATCH.

Spiced Potatoes.

Chop fine twelve cold boiled potatoes, season highly with pepper, salt, butter, and chopped parsley; take one and a half pints new milk, take out a little and rub into it one teaspoonful of cornstarch, and a tablespoonful of butter; heat *all* the liquid, and while doing this prepare enough dry bread-crumbs (seasoned with a taste of cayenne pepper) to cover the top of the dish the potatoes will be baked in, pour the liquid hot on the potatoes, cover the dish with the crumbs, add a few little bits of butter on top to keep them moist, and bake in a hot oven about fifteen minutes.

MRS. SARAH R. BOWDITCH.

Baked Tomatoes.

Select enough medium-sized, perfectly ripe tomatoes to fill a deep baking tin; peel them, scoop out the stem end—placing them in the tin this side up; fill the place with a small piece of nice butter, and cover so thickly with sugar their color is hidden. Bake in a good oven two and a half or three hours, being careful not to burn; when half cooked turn them over. If rightly baked, the tomatoes, when done, will be imbedded in a rich, luscious jelly. If you do not succeed the first time, try again; they are worth the trouble. Remember not to be sparing of sugar.

MATILDA JOSLYN GAGE.

Tomato Bisque.

One quart water, one quart milk, one quart can tomatoes, one teaspoonful soda, two tablespoodfuls cornstarch; cook the tomatoes in the water half an hour, then add soda, then milk which should be hot, cornstarch, a piece of butter half as large as an egg, salt and pepper to taste, strain and serve. I prefer to cook in porcelain.

MRS. B. M. NICHOLS.

Egg Tomato.

Cut up one large tomato in a saucepan over the fire, when it begins to stew break an egg into it, and stir the mixture until the egg is sufficiently cooked; add salt, and, if possible, a spoonful of cream. Canned tomatoes can be used.

ABBY MORTON DIAZ.

SALADS, PICKLES, ETC.

Catsup.

Boil half a bushel of tomatoes, strain through a sieve, add a quart of vinegar, half a pint of salt, an ounce of cloves, two tablespoonfuls of allspice, two teaspoonfuls of cayenne, and one of black pepper. Boil three hours; bottle it air-tight.

MRS. E. J. HARDING.

Plum Catsup.

Boil the plums, skins and all, with a little water, and, when soft, strain through a colander, pressing the pulp through. To five pounds of pulp and juice add three pounds of light brown sugar, one pint best cider-vinegar, one salt-spoonful of black pepper (use cayenne if you prefer), one tablespoonful each of salt, ground cinnamon, allspice, and mace, two teaspoonfuls of cloves (ground). Boil twenty minutes.

MARY F. DANIELL.

Tomato Catsup.

One peck tomatoes in four quarts water. Add nine tablespoonfuls sugar, three tablespoonfuls mustard, one tablespoonful mace, one tablespoonful allspice, two table-spoonfuls cloves, one tablespoonful cayenne pepper, and one pint vinegar. Let it scald a long time.

MRS. C. R. ABBOT.

Tomato Catsup.

Take a half-bushel ripe tomatoes; wash and mash them, and add one teacupful of salt; let them stand over night; add six large onions; put it in a kettle over a slow fire and cook until the onions are soft; when cool, rub it through a sieve. To nine quarts of the liquid add one ounce of cinnamon, one ounce of nutmeg, one-fourth ounce of cayenne pepper, one pint of vinegar (very scant). Always taste of it, and, if needed, add more salt. Let the liquid boil until sufficiently thick, and add the spices and vinegar a few minutes before you take it from the fire. Cloves and allspice can be added if you choose. Bottle when it is cold and seal with wax. MRS. RUTH F. ELWELL.

Tomato Catsup.

One peck tomatoes, quarter cup salt, one teaspoonful red pepper, two teaspoonfuls ground mace, two teaspoonfuls allspice, two teaspoonfuls cloves, two onions, one and one-half pints vinegar. Use *whole* spices, tie in a bag, boil with the tomatoes. After adding vinegar boil till reduced one-half. MRS. M. F. CURTISS.

Mrs. Fogg's Chopped Pickle.

One peck of green, and one peck of ripe tomatoes, one-half head of white cabbage, one-half pound of celery, three onions, four green peppers. Chop fine and drain off the water, then cover with salt and let the whole stand twenty-four hours; drain off the brine and cover with cider-vinegar; add one pound of brown sugar and cook gently three hours, then add one-third cupful grated horseradish, one tablespoonful white mustard-seed, one teaspoonful each of ground cloves, allspice, ginger and mustard. Mix all wel together. MRS. S. C. VOGL.

Tomato Catsup.

Take a half bushel of ripe tomatoes; press through a sieve until you have all the pulp; put the pulp into a porcelain kettle, and when it begins to boil add one-half teacupful of salt, one ounce of whole cloves, one ounce of grated nutmeg, one ounce mace pounded fine, half teaspoonful of cayenne pepper (more cayenne pepper if preferred), one quart of good vinegar. Boil one and a half hours. When cold, bottle, and stop tight. MRS. OLIVER AMES.

Cold Tomato Catsup.

Cut in two a peck of peeled ripe tomatoes; let the seeds and juice drop from them as you cut them; put the pulp in an earthen jar, then add the following: Three pints cider-vinegar, two onions and two green peppers, chopped fine, one teacupful black and white mustard seeds mixed, one teacupful of nasturtium seeds, one cupful of salt, one cupful of sugar, one teaspoonful each of ground cloves and mace, two teaspoonfuls of cinnamon, one teaspoonful ground black pepper, one ounce celery seeds, three celery stalks, three red peppers without the seeds, cut fine, one teacupful grated horse radish. This sauce is fit for use in two days, and will keep all winter. HELEN V. AUSTIN.

Centennial Pepper Hash.

Ten quarts chopped cabbage, one-half cupful of sugar, a handful of salt; squeeze out the juice, and add chopped peppers to suit the taste, one teacupful of white mustard seed, and one of celery seed or chopped celery, add vinegar to cover. MISS M. A. HILL.

Chow-Chow.

One peck green tomatoes, two dozen onions, one quart green peppers, two cupfuls of sugar, handful of cloves, also of allspice; slice the tomatoes and onions, scatter salt over them and allow them to stand over night. In the morning pour off the water and chop fine, then chop in the peppers, and add the sugar and spice, boil until well done (probably three hours) after having covered with vinegar.

REBECCA HOWLAND.

Cucumber Catsup.

Three dozen full grown but green cucumbers and eight white onions, peel and chop them as fine as possible, sprinkle on three-quarters of a pint of fine table salt; put the whole into a sieve and let it drain over night or twelve hours, then add one-half cupful of ground mustard, and one-half cupful of black pepper, and mix thoroughly. Put into glass jars or bottles, and cover with the strongest vinegar. In three or four days it is ready for use. Keep in a cool place; it sealed air-tight it will keep indefinitely.

MRS. ALLIE E. WHITAKER.

Piccalilli.

Slice green tomatoes over night and cover well with salt. Next morning strain off all the liquor and chop fine. To two quarts of the chopped tomatoes put over the fire one quart of vinegar, one pint of water, a cupful of sugar and a quarter of a pound of white mustard seed. Boil twenty minutes; pour in the chopped tomatoes and cook two hours, then add half of a *small* head of cabbage, six or eight onions (according to size), and five or six peppers (part of them ripened), all well chopped. Cook another hour, adding water and vinegar as needed to keep the mixture wet, and sugar as it suits the taste. Put up in sealed jars.

MRS. J. W. GUITEAU.

Piccalilli.

One peck of green tomatoes, chopped and drained, four bell-peppers, four onions, chopped fine, two tablespoonfuls of whole cloves, two tablespoonfuls of allspice, four tablespoonfuls white mustard seed, one tablespoonful of ground mustard, four teacupfuls of sugar, half teacupful of salt. Cover with cider vinegar and boil twenty minutes.

H. A. FOSTER.

Sweet Pickled Cabbage.

Cut a head of cabbage into halves or quarters; after trimming away the finer outside part (which may be used for slaw) boil the heart and stem part of the leaves left in clear water and a little salt till quite tender. Drain well for five or six hours or over night; then to one pint of vinegar add a coffee-cupful of sugar, with whole spices and stick cinnamon to taste (let the cinnamon preponderate). Put cabbage in jar, pour vinegar and spices over while boiling-hot; as soon as cold it will be ready for use; will keep a fortnight. If preferred, the cabbage heart can be left whole, the leaves trimmed till within a couple of inches of it all round, when it is very ornamental for lunch party or *festival* supper.

LOUISE V. BOYD

Fruit Pickle.

For eight pounds of fruit, take four pounds brown sugar, two quarts vinegar, one ounce cinnamon, one ounce cloves or allspice. Let the liquid simmer about twenty minutes, then pour hot on the fruit. MRS. S. W. FULLER.

Pickled Peaches or Pears.

Fourteen pounds fruit, seven pounds brown sugar, a little less than two quarts of vinegar, and a half ounce whole cloves, allspice and cinnamon; tie the spices in linen bags (the cloves may be stuck in the fruit, if preferred). Make the syrup and boil the fruit until tender; pour over the syrup. When cool seal paper over the jar.

MRS. M. F. WALLING.

Pickled Pineapple.

Select the pineapples in June for this purpose. Slice the apple about a quarter of an inch thick, and after paring cut in pieces towards the heart, not using the heart, it is so hard. To every seven pounds of fruit allow three pounds of sugar, one ounce stick cinnamon, one ounce of whole cloves, one quart of vinegar. Put the fruit with the spices into a large earthen crock and mix well. Boil the vinegar and sugar (taking off the scum that rises) and pour it boiling hot over the fruit. The liquor must be drained off and scalded at least six times, but without skimming again, as the spicy flavor would be diminished. Allow three days between each scalding, and put away well covered for winter's use. Pickled peaches are done the same way, only not cut. MRS. OLIVER AMES.

Shaker Pickles.

Take half-grown cucumbers, wash and pack them in jars. Make a pickle of salt and water strong enough to bear up an egg, put a piece of alum in it, as large as a nutmeg to a gallon of pickle; boil and skim and pour hot over the cucumbers; let it remain till cool, and pour off; boil as much vinegar, with spices to suit the taste, as will cover the cucumbers, and pour on while hot. Will be ready for use in two days, and will keep for two years.

HELEN V. AUSTIN.

Tomato Pickle.

One peck green tomatoes and one peck onions; slice them and put over them one cupful of salt; let them stand twenty-four hours. Drain the salt and water off, and boil two or three hours (till tender) in one gallon vinegar; drain off vinegar, and add to the tomato one ounce celery seed. Then boil in the vinegar, one-half ounce mace, one-half ounce cloves, one-half ounce cinnamon, one tablespoonful pepper, one-half cupful white mustard seed, two pounds *dark* brown sugar; pour this over the tomato, onions and celery seed, previously well mixed in a jar.

HARRIET LEMIST.

Green Tomato Pickle, alias Piccalilli.

To one peck of green tomatoes finely sliced add one cupful of salt, sprinkled through the layers of fruit; let this stand over night, and in the morning thoroughly drain off all the liquor. To the above add one-quarter pound mustard seed, one ounce whole cloves, one ounce whole allspice, two cupfuls brown sugar, and two quarts pure cider vinegar. Boil *slowly* for an hour and a half.

SARAH E. M. KINGSBURY.

Salad Dressing.

To three eggs beaten thoroughly add one tablespoonful each of salt, mustard and sugar; to this mixture add one cup milk, one cup vinegar, one tablespoonful butter, and set over the fire, *being careful that it shall only reach the boiling point.* Take off and beat a little and set away to cool.

MRS. S. H. RICHARDS.

Salad Dressing.

For eight pounds of lobster take the yolks of eight boiled eggs, rub them smooth, add eight teaspoonfuls mixed mustard, one teaspoonful salt, one teaspoonful butter, one teaspoonful sugar, and a wine-glass of vinegar.

MRS. RUTH F. ELWELL.

Salad Dressing.

One egg, one teaspoonful sugar, one of butter, one of salt, one of mustard, half cup weak vinegar; mix well together in a bowl; set the bowl in a pail of hot water and stir until it thickens like cold custard.

MRS. M. A. EVERETT.

Salad Dressing.

Four eggs, two teaspoonfuls of mustard, one of salt, two of sugar, a little red pepper, one cup of cider vinegar, butter size of an egg. Make this over the teakettle. Beat mustard, salt, sugar and pepper with the eggs, and stir in the vinegar when just cool enough not to curdle eggs; add the butter as you take it off. Will keep six weeks in cold weather, and three in warm, if in ice-chest.

JANE HOSMER.

Newburyport
Housekeeper's Receipt for Chicken Salad.

The meat of one chicken and one turkey cut in pieces not too small; four small heads of celery; or about equal parts of meat and celery, the latter cut in smaller pieces.

FOR DRESSING. The yolks of eight eggs, stir with silver fork in shallow dish one way, dropping in a few drops of salad oil at a time as you stir; use a small flask of the oil and keep on stirring until very thick; add a heaping teaspoonful and a half of mustard dissolved in two of vinegar, and salt to taste. Mix your celery, turkey and chicken, and taking a little of the dressing thin it with vinegar and mix with your salad, which arrange on a dish, garnishing with celery leaves, and when ready to serve put the dressing, thinned a very little with vinegar, over it—do not, however, do this until just as it goes to the table.

MRS. ELLEN W. E. PARTON.

Lobster Salad.

Chop the meat of two small lobsters and a head of lettuce, mix well together; add the whites of two hard-boiled eggs chopped; mix the yolks of the eggs, one tablespoonful of mustard, one of olive oil, two of vinegar, one of white sugar, a teaspoonful of salt, a little pepper, and the yolks of two raw eggs; pour over, or mix with the lobster.

SARAH F. SARGENT.

Mayonnaise Salad.

One tablespoonful mustard, one tablespoonful sugar, one-tenth teaspoonful cayenne pepper, one teaspoonful salt, the yolks of three un-cooked eggs, juice of half a lemon, one-quarter cupful vinegar, one pint oil, one cupful whipped cream. Beat the yolks and dry ingredients until very light and thick, with either a silver or wooden spoon, or, better still, with a Dover egg beater of second size; the bowl in which the dressing is made should be set in a pan of ice water during the beating; add a *few* drops of oil at a time until the dressing is very thick and rather hard; after it has reached this stage the oil can be added more rapidly; when it gets so thick that the beater turns hard, add a little vinegar — when the last of the oil and vinegar have been added it should be very thick. Now add lemon juice and whipped cream, and place on ice for a few hours, unless you are ready to use it. The cream, though a great improvement, may be omitted without injury.

SARAH E. M. KINGSBURY.

Tomato Salad.

Choose large, smooth tomatoes, remove the skins without tearing the substance by pouring over them hot water; with the curved end of a paring knife remove from the stem end of each tomato the hard portion, thus forming a hollow for receiving the dressing; put the tomatoes thus prepared on ice. With any of the standard dressings (I prefer Durkee's) as a basis, prepare dressing as follows: to one-half pint of this dressing thoroughly beaten to a froth, add two tablespoonfuls of thick, sweet cream, also thoroughly beaten; heat this in a saucepan and when at boiling heat add two teaspoonfuls of gelatine (Coxe's), previously dissolved in just enough cold water to cover it; let this come to a boil and boil three minutes; that it may be quite smooth it should be stirred constantly and rapidly. The dressing should be poured from the saucepan into an earthen bowl; when it is cold, take the tomato from the ice and fill the stem end (hollowed as above indicated) with the dressing. Served, each on a lettuce leaf, in separate small plates, this salad is as decorative to the table as it is agreeable to the palate.

MAY WRIGHT SEWALL.

Chili Sauce.

Chop eighteen ripe tomatoes, one onion, three peppers; add one cup of sugar, two and one-half cups of vinegar, one teaspoonful of salt and of each kind of spice.

MRS. MARY F. CROWELL.

Chili Sauce.

Twenty-four ripe tomatoes, twelve green peppers, six onions; chop onions and peppers together; four tablespoonfuls salt, four tablespoonfuls sugar. Boil together one hour, add six cupfuls vinegar; boil another hour.

MRS. M. F. CURTISS.

Chili Sauce.

Six ripe tomatoes, one large onion, chopped, two peppers chopped fine, one and a half teacups vinegar, one tablespoonful brown sugar, one tablespoonful salt. Cook one and a-half hours slowly. HULDA B. LOUD.

Pepper Sauce.

One pint of mango-peppers, chopped, one quart of cabbage, chopped, enough of mixed spices to flavor. Scald vinegar enough to cover. Can in cans or bottles.

MISS M. L. MORELAND.

PUDDINGS.

Apple Batter Pudding.

Three eggs beaten very light, one pint milk, half teaspoonful salt, one dozen apples chopped fine and added to batter. Bake in cake pans half hour. Serve on platter with cold hard sauce. Mrs. Benj. F. Pitman.

Apple Dumplings.

Two cups sifted flour, two tablespoonfuls lard, one tablespoonful baking powder, half teasponful salt. Peel and core six fair tart apples, fill the holes with sugar, roll each in the crust, and place in an earthen pudding dish. Cover lightly with sugar dotted with small pieces of butter, add half pint of water, cover aud bake two hours. Uncover a short time before taking from the oven to brown.

Mrs. B. M. Nichols.

Apple Pudding.

Six large apples, chopped fine, two tablespoonfuls butter, one small cup sugar. Put in a pudding-dish a layer of grated bread crumbs one inch deep, then a layer of apple, bits of butter, sugar and nutmeg; repeat. Finally pour on a teacup of cold water. Bake half an hour.

Josephine P. Holland.

Sago and Apple Pudding.

Take a pudding dish holding two quarts and fill with sour apple quarters. Sprinkle four large spoonfuls of best sago over them and fill with hot water. Bake an hour, or until the apple is done. It is better to prepare the pudding two or three hours before baking. Eat with sugar and milk.

MRS. SARA T. L. ROBINSON.

Apple Snow Pudding.

Make a custard of one pint rich milk, two tablespoonfuls cornstarch, yolks of two eggs, two tablespoonfuls of sugar and a little salt. Flavor with vanilla when nearly cool; pour into a glass dish; bake six apples, remove the skin and cores, add to the pulp one cup pulverized sugar and the whites of two eggs. Beat till very light and stiff; pour over the custard and add candied cherries or bits of jelly.

MISS E. C. ELDER.

Apple Folly.

Pare, core and bake four apples, add one cup sugar and the white of one egg and beat all together half an hour.

SAUCE.—One pint milk, one egg with the yolk of the one used in the pudding, one and a half teaspoonfuls cornstarch. Put on the stove and stir till it thickens. Flavor with vanilla.

HULDA B. LOUD.

Baked Pie-Plant.

Cut two pounds of pie-plant into a pudding-dish, sprinkle over it half a cupful of sugar and two tablespoonfuls of flour, or what is better, half a cupful of rolled bread-crumbs, add water until the pie-plant is two-thirds covered, bake in a quick oven thirty or forty minutes. This method of preparing pie-plant removes the medicinal taste, and makes an acceptable spring dish.

ALICE B. STOCKHAM, M. D.

Berry Pudding.

Slice and spread with butter a loaf of baker's bread, arrange these slices in a pudding-dish and pour over them hot berry sauce; set away to cool. To be eaten with a hot sauce.

S. LOUISE SIMONDS.

Last Century Blackberry Pudding.

Quart of molasses, teaspoonful each of allspice, cloves and cinnamon, same of salt, three pints of blackberries, two eggs, two teaspoonfuls of baking powder, flour enough to make the spoon stand in the batter; stir the blackberries in the last thing, boil in a bag four to five hours. Can be made just as well without eggs. Blackberries or huckleberries may be used, and, in winter, the same receipt makes an excellent cheap plum pudding.

MRS. LILLIE DEVEREUX BLAKE.

Strawberry or Blueberry Pudding.

Toast slices of bread and place in the dish from which they are to be served; over each slice pour enough canned strawberries or blueberries, thoroughly heated, to soften the bread. Serve hot. This is a delicious pudding, both easily and quickly made. When fresh fruit is used it should be stewed with enough sugar to sweeten it.

JENNIE W. SMITH.

Dark Bread Pudding.

Stale loaf of baker's bread, sliced and buttered; lay in deep pudding-dish with raisins between each slice; cover with a custard made with two quarts of milk and from four to six eggs; dark spices to taste; stand half an hour before baking eight or ten hours in slow oven. Slice cold, and serve with hot sauce.

MRS. H. A. FOSTER.

Steamed Bread Pudding.

Two cupfuls chopped bread, one cupful chopped raisins, half a cupful milk with half a teaspoonful of soda dissolved in it, half a cupful molasses, one egg, one teaspoonful mace, a small half-teaspoonful of all kinds of spice, and a little salt; steam three hours.

SAUCE.—Half a cupful of butter, one cupful sugar, and one cupful of boiling water. SARA A. LOUD.

Chocolate Pudding.

Scald one quart of milk, pour it over four tablespoonfuls grated chocolate, let it stand fifteen minutes to cool; beat together yolks of six eggs, two whites, five tablespoonfuls sugar, one teaspoonful cornstarch, half teaspoonful vanilla; then stir it into the milk and chocolate. Bake in the dish you serve. Beat the remaining whites, add four tablespoonfuls sugar; frost and brown.

MRS. ALICE M. SOUTHWICK.

Nice Cottage Pudding.

One cup sugar, three tablespoonfuls melted butter, one egg, one cup milk, two heaping cups flour, one teaspoonful soda, two teaspoonfuls cream-tartar.

L. F. T. BARNARD.

Plain Cottage Pudding.

One pint flour, one cup milk, one egg, two teaspoonfuls cream-tartar, one teaspoonful soda. Bake twenty minutes, or half hour. L. F. T. BARNARD.

Cracked Wheat Pudding.

In a deep pudding-dish put layers of cold cracked wheat and tart apples sliced thin, with two tablespoonfuls sugar and one tablespoonful raisins. Fill the dish, have the wheat last, and add a cup of cold water. Bake two hours.

CORA L. STOCKHAM.

Delicious Pudding.

Four eggs, yolks beaten separately with four tablespoonfuls flour until very light. Add half teaspoonful of salt, and mix gradually one quart of rich milk with this batter. Beat the whites of the eggs to a froth and stir in last. Bake in a quick oven twenty minutes. It is best to set your pudding dish in a pan of hot water while baking.

MISS A. E. NEWELL.

Egg Sauce for Pudding.

One cupful of sugar, one spoonful of flour stirred up with a little cold water, one tablespoonful of butter, one pint of boiling water. Boil up once and then pour on to a well-beaten egg. MRS. B. J. STONE.

Fruit Sauce.

Boil the juice of any acid fruit, adding an equal part of water. To one pint put one tablespoonful sugar and one teaspoonful cornstarch. This makes a clear juice about the consistence of syrup, and is very desirable to eat with wheat, mush, gems, griddle cakes and plain puddings. Jellies and jams can be made into fruit sauce by adding four parts of water and thickening. Will not require sugar. These are valuable sauces for invalids and children. Once learning how delicious they are, persons in health will demand them. In many of the small fruits the seeds are very objectionable. This method of using the fruit obviates that.

ALICE B. STOCKHOLM, M. D.

Fruit Pudding.

Put into a nappy your fruit,—apples, peaches, canned cherries or blueberries, taking care with canned fruit not to put in much juice—(that can be served with the pudding in a separate dish). Have the fruit sweetened, then cover an inch deep with finely grated bread-crumbs. Bread which is not dried grates most easily. Eat with cream.

MRS. EMMA E. FOSTER.

German Puffs.

Six eggs, leaving out the whites of three for sauce, five tablespoonfuls flour, one tablespoonful melted butter, one pint milk. Bake in cups thirty minutes; makes nine puffs.

SAUCE.—Beat the whites to a stiff froth. To one cup of sugar add the juice of one lemon or of two oranges. Add it to the whites and pour over the puffs just before sending to the table. HARRIET LEMIST.

Graham Pudding.

Two cups graham flour, one cup sweet milk, one cup molasses, one cup raisins (with a few slices of citron), one teaspoonful *each* of salt, soda and cinnamon, half teaspoonfull cloves. Steam an hour and a half in a basin or cake-pan. Serve hot with any desirable pudding sauce.

DR. L. G. BEDELL.

Steamed Graham Pudding.

Three cupfuls graham, one cupful sour milk (or better still, half a cupful or more of sour cream), one cupful molasses, two teaspoonfuls soda and salt. Fruit of any kind, fresh or dried, to suit the taste. Eat hot with cream. Steam at least three hours. MRS. EMMA E. FOSTER.

Honey Comb Pudding.

One-half cupful of flour, one-half cupful of sugar, one-half cupful of milk, one-half cupful of butter, one-half pint New Orleans molasses, four eggs, one teaspoonful soda. Mix flour, sugar, butter, and milk warm enough to dissolve the butter, all together; add the beaten eggs, and last, the molasses and soda beaten to a froth. Bake half an hour. Serve with cream, or foamy sauce. MRS. C. S. LINCOLN.

Indian Pudding.

One cupful corn meal, one cupful molasses, one teaspoonful salt, five pints milk, one tablespoonful butter, one egg. Mix the corn meal, molasses and salt. Cook it in one quart of boiling milk five minutes. Add one tablespoonful of butter and one beaten egg. Put in a deep, well-buttered dish. When in the edge of the oven, pour in carefully three pints of cold milk. Bake seven or eight hours.

MRS. L. W. JONES.

Indian Pudding.

Four tablespoonfuls meal mixed with four tablespoonfuls molasses and one teaspoonful salt. Pour over this one quart scalding milk, mix well, add one cup cold milk, two-thirds cup raisins, two-thirds cup flour, after sifting. Bake two hours.

MRS. M. A. EVERETT.

Old-Time Baked Indian Pudding.

Three pints of sweet milk, two large iron spoonfuls of yellow cornmeal, one small egg, one iron spoonful of molasses, three-fourths cup of sugar, heaped teaspoonful of ginger, level teaspoonful of cinnamon, one-third of a small nutmeg, and one-half a teacupful thick, sour cream. Put half the milk over the fire with a sprinkling of salt; as soon as it comes to a boil scatter the meal quickly and evenly in by hand. Remove immediately from the fire to a dish, stir in the cold milk, the egg well-beaten, the spices, sweetening, and sour cream. Bake three hours, having a hot oven the first half hour, a moderate one the remainder of the time. Eat with sweet cream. If rightly made and rightly baked, this pudding is delicious, but four things must be remembered as requisite: First, the pudding must be thin enough to run when put in the oven. Second, the egg must be small, or if large, but two-thirds used for a pudding of the above size. Third, the sour cream must not be omitted (but in case one has no cream, the same quantity of sour milk with a piece of butter the size of a small butternut can be substituted). Fourth, the baking must be especially attended to. Many a good receipt is ruined in the cooking, but if the directions are carefully followed, this pudding will be quavery when done, and if any is left, a jelly when cold. Use no sauce, but sweet cream or butter.

MATILDA JOSLYN GAGE.

New Jersey Indian Pudding.

Add to one quart boiling milk, salted to taste, enough sifted yellow Indian meal to make a moderately stiff batter or "mush." Keep boiling ten or fifteen minutes, taking care to prevent scorching; when partially cool add two eggs (not beaten), one at a time, piece of butter size of egg, more if liked. Sweeten to taste with sugar; half a cupful of molasses or less added to make it brown nicely is liked by some. Bake slowly one hour in an ordinary sized pudding dish, or less in two smaller ones. MRS. H. O. HAWKINS.

Kiss Pudding.

Place one pint of milk in a double boiler, when nearly boiling add the yolks of two eggs, two tablespoonfuls sugar and two teaspoonfuls cornstarch beaten together. Stir until thickened, remove from the stove, flavor and turn into dish to serve. Beat the whites of two eggs and a pinch of salt to a stiff froth, add three teaspoonfuls of sugar, turn it on the pudding and slightly brown in oven.

MRS. H. ANDREWS.

Orange Souffle.

Peel and slice six oranges, removing the seeds. Put in a high glass dish a layer of oranges, then one of sugar, and so on until all the oranges are used. Let it stand two hours. Take yolks of three eggs, one pint of milk, and sugar to the taste. Flavor with the grated orange peel, and make a soft boiled custard aud pour over the oranges when cool enough not to break the dish. Beat the whites of the eggs to a stiff froth, stir in sugar and pour over the pudding and it is done.

MARY A. LIVERMORE.

Plum Duff.—Navy Receipt.

One cupful molasses, one cupful sweet milk, one large cupful suet chopped fine, one teaspoonful of soda, one-half teaspoonful of cream-tartar, two eggs, one large cupful of chopped raisins; nutmeg, clove, allspice, cinnamon ground, a heaping teaspoonful all mixed together; three good cupfuls of flour, salt sufficient to season. Steam in pudding pan or mould at least four hours, and seven is not too much. Eat with cold sauce. This can be made the day before eating advantageously, and heated, and is an excellent, hearty plum pudding, not rich. Well tested receipt.

MRS. SARAH R. MAY.

Plum Pudding.

Over one teacupful of powdered cracker and one quarter pound of beef suet, chopped fine, pour one quart of boiling milk. When cool add one teacupful boiled raisins, a little salt, one-half teacupful of sugar and five eggs, leaving out one white for the sauce. Bake in pudding dish one hour and stir while baking.

SAUCE for the above.—Beat one and a quarter cupfuls of powdered sugar and one quarter pound of butter to a cream; then add the white of one egg, well beaten, and flavor to taste. Put in a glass dish and grate nutmeg over the top.

MRS. ELIZA B. BURGESS.

Belle's Favorite Pudding.

Four tablespoonfuls pearl tapioca, twelve apples pared and quartered. Soak the tapioca in cold water till clear, add the apples and a little salt, together with any flavoring which suits the taste. Bake one-half hour in a quick oven, and serve with cream. MRS. MARY CROSS HARRIS.

Every-Day Plum Pudding.

One cup molasses, one teaspoonful soda, one cup hot water, one teaspoonful salt, one cup suet, chopped fine, one teaspoonful clove, three cups flour, one teaspoonful cinnamon, two cups raisins, cut up fine (not chopped), half cup sliced citron. Dissolve the soda in the hot water, add the molasses, suet, salt and spices; stir in the fruit with the flour, last. Boil or steam four hours. Eat with hard sauce.

HARRIET H. ROBINSON.

Thanksgiving Plum Pudding.

Fifteen crackers, five eggs, one pound raisins, two and one-halt cups of sugar, four sweet apples, one cup of suet, two quarts sweet milk, salt and mace to taste. Pound the crackers, and soak in one quart of milk over night. In the morning add the other quart, with the rest of the ingredients; stone the raisins. Bake four hours and steam four hours. (The steaming may be omitted, and a longer bake given if preferred.)

MISS M. A. HILL.

Popover Pudding.

One pint of milk, one pint of flour, three eggs, and bake one hour. To be eaten with sauce.

MISS A. E. NEWELL.

Pudding.

Three cups Indian meal, three cups flour, three eggs, three tablespoonfuls sugar, one pint milk; salt to taste. Scald the meal with half the milk, and let it cool before adding other ingredients. Bake an hour. Stir once after it has begun to cook.

REV. LOUISE S. BAKER.

Pudding Sauce.

Boil half a pint of water, salt it and thicken with two tablespoonfuls of flour mixed smooth with as little cold water as possible. Stir till thick and smooth; stir in a teaspoonful of butter and let it cool. Beat the yolks of two eggs and a cup of sugar to a cream and stir into the cold paste. Flavor to suit the taste, stir till smooth, add the whites of the eggs beaten to a stiff froth, and beat all to a foam.

MRS. L. W. JONES.

Pudding Sauce.

One cupful sugar, one tablespoonful butter, one egg, one cupful water. Stir together and boil. Add nutmeg.

MISS L. F. S. BARNARD.

Rice Pudding.

Take two quarts of good new milk, put in a cupful of rice, one of raisins, one of sugar and a tablespoonful a little heaped of butter. Put on back of stove and cook slowly six hours.

MRS. D. W. GAGE.

Sponge Pudding.

Quarter cup of sugar, quarter cup of butter, half cup of flour, yolks of five eggs, one pint of milk, boiled, whites of five eggs. Mix the sugar and flour, wet with a little cold milk, and stir into the boiling milk; cook until it thickens and is smooth; add the butter, and when well mixed stir it into the well-beaten yolks of the eggs, then add the whites beaten stiff. Bake in cups, or in a shallow dish, or in paper cases, in a hot oven. Place the dish in a pan of hot water while in the oven. Serve with creamy sauce.

MRS. C. S. LINCOLN.

Steamed Pudding.

One and a-quarter cups flour, half a pint of milk, sweet or sour, half a pint molasses, one-third cup butter, one teaspoonful soda if sweet milk is used, and two, if sour; half a teaspoonful salt; fresh or dried fruit. Steam three hours.

MRS. EMMA E. FOSTER.

Frosted Tapioca Pudding.

Three tablespoonfuls of pearl tapioca, soaked over night in one cup of water; add one quart of milk; when it boils, add the yolks of four eggs and one cup of sugar, beaten together. Flavor and salt to the taste.

FROSTING.—Beat the whites to a froth, add four tablespoonfuls of sugar gradually, and beat until you can turn over the bowl without the liquid running out. Drop it on the pudding in lumps, and brown in the oven.

MRS. JOSIE CURRIER.

Indian Tapioca Pudding.

Two tablespoonfuls tapioca soaked over night, three tablespoonfuls Indian meal, one quart milk, one cupful cold water, two-thirds cupful molasses, a little salt and a small piece of butter. Mix tapioca, meal, molasses, salt and butter together and pour over it a quart of boiling milk. Stir thoroughly and just before putting in the oven add the cupful of cold water. Bake two hours, stirring once or twice after it begins to bake. MRS. WILLIAM C. COLLAR.

Snow Pudding.

Pour over three tablespoonfuls of corn starch dissolved in a little cold water one pint of boiling water. Add the whites of three eggs well beaten and steam twenty minutes.

SAUCE.—After scalding one cupful of milk stir in the yolks of three eggs beaten with a cupful of sugar and a small piece of butter; when cold flavor with vanilla.

MRS. AUGUSTA RICH.

Snow Pudding.

Soak one-fourth box gelatine in one-fourth cupful cold water until soft, then add one cupful boiling water, one cupful sugar and one-fourth cupful lemon juice. Strain into a bowl and set on ice. Beat the whites of three eggs to a stiff froth, and when the jelly begins to thicken add the beaten whites and beat all together till very light. Pour into a mould. Make a boiled custard, using the yolks of the three eggs, three tablespoonfuls of sugar, one pint of milk and a little salt. When the jelly is moulded put it in the dish from which it is to be served and pour the custard over it. Garnish with sliced fruits or blocks of jelly.

MISS E. C. ELDER.

Excellent Snow Pudding.

Half a box gelatine, half a pint cold water; let it soak half an hour, then add half a pint of boiling water. When cool, add the whites of three eggs, two cups sugar, juice of two lemons, and beat the whole well half an hour or more. Set away to cool in a mould. Make a boiled custard of the yolks of the eggs, and one and one-half pints milk; sugar to taste. Serve the solid part floating in the custard, with whipped cream poured over the top.

MRS. FORREST W. FORBES.

Apple Truffles.

One dozen juicy apples of fine flavor, two cups white sugar, one scant quart of milk, four eggs, one pint of sweet cream, two tablespoonfuls of powdered sugar, juice and grated rind of one lemon. Slice the apples and put them into a glass jar; cover closely, set in warm water, bring to a boil, and cook until tender and clear, then beat to a pulp, sweeten with one cup of sugar and the rind and juice of the lemon. Put the mixture into a glass dish, and set in a cold place. Make a custard of one quart of milk, the four eggs, well beaten, and one cup of sugar. Put into a double boiler and cook until it thickens. When cold, flavor to taste with vanilla. Pour this over the apple, which must be ice-cold and stiff to prevent its rising. Whip the cream to a stiff froth (with the two tablespoonfuls of sugar), and pile over all.

MRS. ALICE A. GEDDES.

Delicate Indian Pudding. (Parloa.)

One quart milk, two heaping tablespoonfuls of Indian meal, four of sugar, one of butter, three eggs, one teaspoonful of salt. Boil milk in double boiler, sprinkle the meal into it, stirring all the while; cook twelve minutes, stirring often. Beat together the eggs, salt, sugar and one-half teaspoonful of ginger. Stir the butter into the meal and milk. Pour this gradually over the egg mixture. Bake slowly one hour. Serve with sauce of heated syrup and butter.

MRS. BENJ. F. PITMAN.

Raspberry Pudding.

One-quarter cupful of butter, one-half cupful of sugar, two cupfuls of jam, six cupfuls of soft bread-crumbs, four eggs. Rub the butter and sugar together; beat the eggs, yolks and whites separately; mash the raspberries, add the whites beaten to a stiff froth; stir all together to a smooth paste; butter a pudding-dish, cover the bottom with a layer of the crumbs, then a layer of the mixture; continue the alternate layers until the dish is full, making the last layer of crumbs; bake one hour in a moderate oven. Serve in the dish in which it is baked. MRS. ALICE A. GEDDES.

Raspberry Sauce.

One-half cupful of butter, one cupful of sugar, one cupful of raspberries, the whites of two eggs. Rub the butter and sugar to a cream, beat the whites of the eggs to a stiff froth, and add to the butter and sugar; mash the raspberries, stir all together until smooth, and serve.

MRS. ALICE A. GEDDES.

PIES.

Pastry.

Lard or butter to be used for pastry should be as hard as possible. It should be cut through the flour with a chopping knife, not rubbed.

Pastry for Apple or Mince Pies.

One cupful lard, one-half cupful butter, two cupfuls flour, a pinch of salt. Cut the lard through the flour, then add ice water enough to mix, using a knife; do not knead; roll out and place small bits of butter over it, dredge with flour, roll up and repeat the process, using up the half cupful of butter. Roll this and bake in a quick oven. Brushing the paste as often as rolled out with the white of an egg assists it to rise in leaves or flakes. The amount of lard and butter can be varied, and the paste made richer or plainer as desired.

Sliced Apple Pie.

Line the pie-plate with a rich paste, sprinkle with sugar, then add a layer of tart apples sliced thin; add a few bits of butter, grated nutmeg, and cover with sugar. Repeat the layer of sliced apples, with the butter, nutmeg and sugar, until the plate is filled. Dredge flour over the whole and cover with the top crust, tucking it under the bottom crust around the edge to prevent the juice from running out. Bake three-quarters of an hour. MARY A. LIVERMORE.

Cream Pie.

Half cup butter, one cup sugar, one egg, yolks of three more, one teaspoonful of cream-tartar, half teaspoonful soda, one and one-half cups flour, half cup milk.

CREAM.—One pint milk, taking out half cup for cake; mix half cup flour with one cup sugar, and stir into the milk while boiling. Boil a few minutes; take from the fire and add the three whites, beaten stiff; vanilla.

SARA A. LOUD.

Cream Pie.

Three eggs, whites and yolks beaten separately, one cupful sugar, one heaping cupful flour, one even teaspoonful of soda dissolved in three tablespoonfuls of milk, two even teaspoonfuls of cream of tartar mixed in the flour, a little salt and nutmeg; split while hot.

CREAM for filling.—One cupful sugar beaten with two eggs, one pint boiling milk, three tablespoonfuls cornstarch wet smooth with cold milk; stir the mixture in the milk. Flavor with lemon or vanilla. Fill while hot.

MRS. W. C. COLLAR.

Currant Pie.

One cupful ripe currants, one cupful sugar and one egg well beaten. Stir in the currants. Bake between two crusts.

MRS. M. J. WILLIS.

Lemon Pie.

Line a pie plate with a rich paste. Grate the rind of one large or two small lemons and then squeeze out the juice. To this add the yolks of two eggs, one cupful of sugar, and finely powdered cracker sufficient to make a tolerably thick batter, beating the whole well together. Pour this into the pie plate and bake. Take the whites of the two eggs and stir in powdered sugar to make a frosting. Now pour this over the pie, set it back in the oven a few moments till well set and of a delicate brown in color, and the pie is done.

MARY A. LIVERMORE.

Lemon Pie.

Three lemons, using the juice and grated rind, three cupfuls of sugar, two heaping teaspoonfuls of flour made into a paste with one cupful of cold water, four or five eggs. Mix the yolks with the paste, lemon and sugar; use the whites for frosting. HELEN B. W. WORTH.

Lemon Pie.

Grated rind and juice of three lemons, half a cupful of melted butter, two tablespoonfuls of cornstarch, two cupfuls of water, three cupfuls of sugar and nine eggs, whites and yolks beaten separately. Stir all together, beating the whites in well the last thing. Bake with one crust.

MRS. M. L. T. HIDDEN.

Mince Pie Meat.

One bowlful of chopped raisins, three bowlfuls of chopped meat, six bowlfuls of chopped sour apples, one bowlful of chopped suet, one bowlful of molasses, one bowlful of sugar, two lemons, chop the peel very fine, four tablespoonfuls of salt, four teaspoonfuls of clove, four teaspoonfuls of allspice, six teaspoonfuls of cinnamon, one teaspoonful of black pepper, one nutmeg, grated. Moisten with the liquor that the meat is boiled in, and boil the whole thoroughly.

MRS. JOSIE CURRIER.

Mince for Pies.

Order the choicest rump-steak, say a piece weighing four and one-half pounds, which will boil down to about two pounds; two pounds of meat, chopped fine, four and one-half pounds of apple, chopped fine, three-quarters pound of suet, chopped fine, four and one-half heaping tablespoonfuls of powdered cinnamon, one heaping tablespoonful of powdered clove, three tablespoonfuls of salt, three and one-half pounds of sugar, three nutmegs, grated fine, three pounds of raisins, boiled until very tender. Mix the mince well, adding a coffee-cup, or more, of the raisin liquor and peach-pickle vinegar, or any choice sweet pickle-juice, until the mince is sufficiently moist. Reserve a few of the boiled raisins to lay over the top of the pies with small bits of butter before putting on the upper crust. More spice can be added to suit the taste, if required. MRS. OLIVER AMES.

Mince-Pie Meat.

Four pounds of meat, after chopping, eight pounds of apples, two pounds of suet boiled one hour, five pounds of sugar, two pounds of raisins, two pounds of currants, one pint molasses, one quart sweet cider, one cupful of salt, two tablespoonfuls of clove, two tablespoonfuls of cinnamon, one large spoonful of allspice, and four nutmegs. Chop half of the raisins, and boil meat without salt. This rule will make forty pies.

MRS. MARCIA E. P. HUNT.

Mince Pie Without Meat or Apple.

Three crackers rolled fine, one cupful molasses, one cupful vinegar, one cupful sugar, one cupful raisins chopped, one-half cupful water, one-half cupful butter, one egg, one teaspoonful each cassia and clove, one nutmeg. This makes three pies. MRS. DR. FLAVEL S. THOMAS, M. S.

Vermont Pumpkin Pie.

Peel and cut your pumpkin into small pieces and put into a kettle with a very little water, cook from six to eight hours, stirring frequently to prevent burning. When done it should be quite dry and of a rich brown color. Rub through a colander. One quart of pumpkin, three pints of rich milk, four eggs, two and a half cupfuls sugar, one tablespoonful ginger. Bake in rather a slow oven until nicely browned.

MRS. M. L. T. HIDDEN.

Squash Pie.

To one pint sifted squash add one quart boiling milk, one egg, two crackers rolled fine, one large cupful sugar, one teaspoonful cornstarch, half teaspoonful cinnamon, half teaspoonful salt and a little nutmeg. MRS. H. ANDREWS.

Washington Pie.

FOR THE CAKE.—Four eggs, one cup of sugar, tablespoonful of butter, one cup of flour, one teaspoonful baking powder.

FOR THE FILLING.—One sour apple, pared and grated, the juice and grated rind of a lemon, one cup of sugar, all boiled together for five minutes.

SARAH F. SARGENT.

Washington Pie.

Half cup of butter, thoroughly beaten with one cup of powdered sugar; add to this two eggs separately beaten, two-thirds cup of sweet milk, two cups St. Louis flour, with scant teaspoonful of cream of tartar and half teaspoonful bread-soda sifted and faithfully mixed in the flour. This is sufficient for two large pies. For filling I use an acid jelly.

SARAH E. M. KINGSBURY.

Washington Pie.

Two-thirds cup of sugar, two tablespoonfuls butter, one egg, two-thirds cup of milk, two teaspoonfuls cream of tartar and one of saleratus; mix a trifle thicker than griddle-cakes.

CREAM to fill Washington or cream pie.—One cup milk, one egg, one half cup sugar, two dessert spoonfuls flour; boil the milk, beat the sugar, egg and flour together, and stir it in the boiling milk until it thickens.

MRS. C. A. SARGENT.

DESSERTS, CREAMS, ETC.

Charlotte Russe.

One pint of cream beaten to a froth, whites of two eggs well beaten, one cupful sugar; flavor with vanilla; add one-fourth box gelatine dissolved in a gill of hot milk.

SPONGE CAKE for the above.—One cupful of sugar, one of flour, three eggs, yolks and whites beaten separately, one-half teaspoonful soda, one teaspoonful of cream of tartar, flavor to taste, bake slowly. Fill the mould with sponge cookies cut to fit, or a thin layer of this cake; a deep tin cake pan will answer. MRS. MARTHA J. WAITE.

Raspberry Cream.

One-half cupful of cream whipped, the white of one egg beaten, two tablespoonfuls of gelatine dissolved in hot water —add slowly to the beaten egg, add the cream, two tablespoonfuls of sugar, two of raspberry jam, beat all together; set in a cool place. Delicious as a dessert, or as a sauce for a cottage pudding. JUDITH WINSOR SMITH.

Bavarian Cream.

Let one quart of milk come to a boil. Beat the yolks of two eggs and two large tablespoonfuls of sugar to a froth; to this add one-third of a box of gelatine dissolved. Stir in the boiling milk and cook a few minutes, and when half cool strain into the dish to be used on the table. Beat the whites of the eggs and add to them two tablespoonfuls of powdered sugar as in frosting. Stir through the milk, set it on ice and let it stand from twelve to twenty hours. Serve with cream and a little jelly or preserve. MARY GAY CAPEN.

Tapioca Cream.

Soak two tablespoonfuls of tapioca over night in just water enough to cover it. Stir the tapioca into one quart of boiled milk; this must just boil, and then stir in the beaten yolks of three eggs and one cupful sugar, and let it boil until it thickens. Flavor slightly with vanilla. After it is poured into the dish beat up the whites of the eggs with a tablespoonful of sugar and put upon the top. To be eaten cold. The custard may be made the day before.

HARRIET LEMIST.

Tapioca Cream.

Soak two tablespoonfuls of tapioca for two hours; boil one quart of milk, and add the tapioca, put in the yolks of three eggs, well beaten, with a cup and a-half of sugar; let this boil up, and set it away to cool. Beat the whites to a froth with a Dover egg-beater, and add to the top as for boiled custard. EMMA M. E. SANBORN, M. D.

Tapioca Cream.

One quart milk, yolks of four eggs, two tablespoonfuls of tapioca, four tablespoonfuls of sugar; soak the tapioca over night, in warm water, heat milk, add tapioca, and let it cook until tapioca rises to the top. Beat the whites of the eggs to a stiff froth, add powdered sugar. Cover the cream and brown it lightly in the oven. Flavor to taste; eat cold. DR. VESTA D. MILLER.

Baked Custard.

Six eggs, one quart milk, four tablespoonfuls sugar; add nutmeg and salt; set the pudding-dish in a pan of water and bake. Test with a broom-splint, and remove from the oven as soon as done, or it will become watery.

MISS L. F. S. BARNARD.

Corn Custard.

Cut corn frcm the cob, mix with it a little milk, two or three eggs, pepper and salt. Bake half an hour. Nice addition to a dinner. Eat with meat. MISS L. A. HATCH.

Plain Custard.

Four tablespoonfuls flour, six tablespoonfuls sugar, one pint milk, and one egg; flavor as you choose. Mix the sugar and flour thoroughly, and when the milk is boiling pour it on the mixture and stir well. Return to the stove, and when the whole is well cooked, put in the beaten yolk as in boiled custard. After two or three minutes of stirring and cooking, take from the stove and beat in the flavoring and the white of the egg which has been beaten to a stiff froth. Pour out and set away to cool.

MRS. EMMA E. FOSTER.

White Custard.

To one quart milk add one cupful granulated sugar; put in a farina boiler and let come to a boil. Pour this over the whites of six eggs, beating all the time; add flavoring. Put in custard cups and bake in a pan with water around the cups to keep them from breaking. A very delicate dessert for invalids. MRS. THOS. S. LYON.

Peach Sherbet.

Two quarts of peaches cut and sugared over night, using two cupfuls of sugar. Put a tablespoonful of gelatine in half a cupful of cold water, let it stand half an hour, then add half a cupful of hot water. Put the peaches through a colander or sieve, add gelatine and a pint of ice water and freeze. MRS. D. P. WASHBURN.

Pineapple Sherbet.

Peel and chop fine a large pineapple, add a large pint of sugar, let it stand over night. Take one tablespoonful of gelatine, just cover with cold water, let it stand until it is soft, add one-half pint of boiling water, when fully dissolved add one-half pint of cold water, strain the gelatine and mix with the pineapple and sugar. Freeze like ice cream.

MRS. M. E. SAMMET.

Nice Table Syrup.

One quart of sugar, two tablespoonfuls of molasses, one quart of water, boiled slowly until reduced to little less than a quart; if boiled longer, it becomes candied.

MISS C. WELLINGTON.

Chocolate Blanc-Mange.

One quart milk, half box Cox's gelatine, one square of Baker's chocolate; scald the milk, add the galatine, dissolved in a little water, and the grated or melted chocolate. Let it scald again, and before it is put in the mould add a little vanilla extract. Make the day before it is wanted, as it takes some time to harden. One square of chocolate makes a very delicate blanc-mange; more may be used if the strong flavor of chocolate is desired.

CORA SCOTT POND.

Ice Cream.

One quart cream, one pint milk, four eggs well beaten, one and a half cupfuls sugar, one tablespoonful vanilla or other flavoring. No cooking is required.

MRS. S. J. G. BECK.

CAKES.

Angels' Cake.

Whites of eleven eggs beaten to a stiff froth, one and one-half cupfuls pulverized sugar, one teaspoonful cream of tartar mixed with one cupful flour; sift the flour three times; flavor with one teaspoonful vanilla (never use lemon). Beat only one way with a spoon or fork, bake forty minutes without moving. MRS. FORREST W. FORBES.

Bangor Cake.

Three and one-half cupfuls of flour, two cupfuls of sugar, one-third cupful of butter, three eggs, one cupful of milk, two teaspoonfuls of best cream of tartar, one teaspoonful of soda. Beat sugar, butter and cream of tartar together, beat the eggs separately, then beat them in the mixture and add the other ingredients. Add the soda dissolved in a little of the milk the last thing before baking. MARY WILLEY.

Berry Cake.

One cupful milk, one egg, one cupful berries, two and one-half cupfuls flour, three tablespoonfuls melted butter, baking powder. Bake in gem pan. MISS A. E. NEWELL.

Bradford Cake.

Whites of three eggs beaten to a stiff froth, add one cupful of sugar and beat five minutes, one-half cupful of butter well beaten, one-half cupful of sweet milk, one teaspoonful of cream of tartar, one-half teaspoonful of soda, two cupfuls of flour. Flavor with vanilla.

Chocolate Frosting.

Grate two squares of chocolate, add one cupful of sugar and four tablespoonfuls of boiling water; place in small stew-pan and let it boil ten minutes. Spread quickly on warm cake. MRS. DR. C. A. CARLTON.

Cake.

Butter size of a small egg, two-thirds of a cupful of sugar, one egg well beaten, one-half cupful of milk, one teaspoonful of cream of tartar and one-half teaspoonful soda. Beat the sugar and butter to a cream, then add the other ingredients. MRS. MARY J. BUCHANAN.

Cakes for Children.

One pint flour, one cupful sugar, half a cupful butter and a pinch of salt. Mix with milk or water, flavor with lemon, rise with baking powder, or cream of tartar and soda, or yeast. Bake in little scallop tins. ABBY MORTON DIAZ.

Caramel Cake.

One cupful sugar, one-half cupful butter, two eggs, one-half cupful milk, one teaspoonful cream of tartar, one-half teaspoonful soda, two cups flour.

CARAMEL for the cake.—Two cupfuls sugar, two-thirds cupful milk, butter size of an egg, boil ten minutes, beat till cold. JERUSHA S. HALL.

Chocolate Cake.

Beat to a cream one and a half cupfuls of sugar with one-half cupful of butter; add three eggs (whites and yolks beaten separately), one-half cupful milk, one-half teaspoonful soda dissolved in the milk; add eight tablespoonfuls grated chocolate, three tablespoonfuls sugar, two tablespoonfuls scalded milk, dissolved together by steaming over a teakettle; then add two cupfuls of flour with one teaspoonful of cream of tartar. Frost the top with white frosting.

MRS. E. E. KELSEY.

Chocolate Cake.

One-half cupful butter, one cupful sugar, two eggs, one-half cupful milk, one and one-half cupfuls flour, one and one-half teaspoonfuls baking powder.

ICEING.—One cupful sugar, five tablespoonfuls of water; boil till it strings. Pour in it one and one-half squares of Baker's chocolate melted; let this boil till it strings, then stir this into the beaten whites of the eggs till it hardens; spread between the layers of cake. MARY GAY CAPEN.

Citron Cake.

One cupful of butter, two cupfuls of sugar, three cupfuls of flour, three tablespoonfuls of molasses, three eggs. two cupfuls of currants, one cupful of citron chopped fine, one cupful of milk, one teaspoonful of cream of tartar, one-half teaspoonful of soda, one teaspoonful each of clove and cinnamon, one-half a nutmeg. SARAH F. SARGENT.

Cocoanut Cake.

One-half cupful of butter, one and one-half cupfuls of powdered sugar, two cupfuls of flour, one-half cupful of milk, whites of five eggs, one teaspoonful cream of tartar, one-half teaspoonful soda, juice of half a lemon. Beat the butter and sugar to a cream, add the lemon juice, then the milk, and later the whites, beaten to a stiff froth, lastly add the flour into which the soda and cream of tartar have been mixed. Bake in two sheets in a moderate oven about half an hour. When a little cool frost with the following

FROSTING.—Whites of three eggs, two large cupfuls of powdered sugar, one-half pound can of cocoanut. Beat the whites to a stiff froth, then add the sugar and cocoanut and juice of half a lemon. Put half the frosting between the sheets and the rest on top. St. Louis flour to be used.

SARAH E. M. KINGSBURY.

Cocoanut Cookies.

Into two and one-half cupfuls of pastry flour, rub with the hands one-half cupful of butter. Add one cupful of sugar, one and three-fourths cupfuls of grated cocoanut (that which comes by the pound is best) and two saltspoonfuls of cream of tartar. Beat one egg and stir in ; dissolve one saltspoonful of soda in boiling water and add, moulding the mixture well together with the hands. If it is not wet enough, add a very little milk or water. The danger is in getting it too wet to roll out well, and probably no moisture will be needed. Roll thin, cut with a doughnut cutter and bake quickly.

MRS. H. R. SHATTUCK.

Soft Molâsses Gingerbread.

One cupful of sugar, one-half cupful of butter, one cupful of milk, one egg, one cupful of molasses, one teaspoonful of soda in the milk, three scant cupfuls of flour, ginger or not as you like. LUCY GODDARD.

Coffee Cakes.

Beat three eggs very light, add two cupfuls brown sugar, one cupful butter, one cupful sweet milk, one teaspoonful soda, two teaspoonfuls cream of tartar. Make a stiff dough by kneading in flour, roll out to about one-half inch thick, sprinkle with powdered sugar and cinnamon, roll up as if for jelly rolls, and cut off slices about half an inch thick, dip in granulated sugar and bake.

MRS. DR. FLAVEL S. THOMAS, M. S.

Cookies.

One cupful sour rich cream, one cupful white sugar, one-half teaspoonful soda, flour enough for a soft dough—only enough to roll out easily; salt and nutmeg if desired.

MRS. SARA T. L. ROBINSON.

Cream Cookies.

Two cupfuls of sugar, one cupful of butter, one-half cupful of sour cream, one level teaspoonful of soda dissolved in a little hot water, enough flour to roll out as soft as possible.

EMILY S. BOUTON.

Fruit Cookies.

One and one-half cupfuls sugar, one-half cupful butter, five tablespoonfuls milk, one teaspoonful soda, spice of all kinds, one cupful currants or raisins, chopped, flour to roll out thin.

MRS. L. W. JONES.

Lep Cookies.

One gallon molasses, two pounds lard, one pound citron, one teacupful each of cinnamon and spice, one-half teacupful cloves, four or six nutmegs, two pounds picked nuts (hickory or pecans), flour to make a stiff dough; roll thin, and bake quickly; ice, and dry well before putting away.

MRS. JESSIE F. A. BANKS.

Molasses Cookies.

Put into a large coffee-cup one teaspoonful of soda, two tablespoonfuls of *hot* water and three tablespoonfuls of melted butter. Fill the cup with molasses, add a little ginger if liked. Two cups are enough for one baking. Mix soft and bake quickly.

LOUISA G. ALDRICH.

Guess Gingerbread.

Lard size of one egg and a half, quarter cup sour milk or buttermilk, three-quarters cupful molasses and half cupful brown sugar, one egg (beaten separately), salt and ginger to taste, *even* teaspoonful soda; flour for soft batter.

J. S. FOSTER.

Molasses Cookies.

One egg, one cup molasses, one-half cup of sugar, one teaspoonful each of salt, soda and ginger; flour enough to roll easily. This receipt calls for neither milk or shortening, and makes very nice cookies. Bake in quick oven.

MRS. ELLIE A. HILL.

New Bedford Cookies.

Two cups of sugar, one cup of sour milk with half a teaspoonful of soda dissolved in it, one cup of butter. Flour to roll not too stiff, and bake quickly.

EMILY A. FIFIELD.

Spiced Cookies.

One cup of sugar, two cups of molasses, two-thirds of a cup of butter, one cup of milk, one teaspoonful of soda, one small teaspoonful of cloves, and one small teaspoonful of cinnamon, two eggs, one-half a nutmeg, and five cups of flour.

LOUISA G. ALDRICH.

Sugar Cookies.

One egg, one cup of sugar, half a cup (scant) of butter, half a cup of milk, nutmeg to taste, two teaspoonfuls of cream of tartar, one of soda. Make soft dough as can be handled; roll thin and bake in quick oven.

MRS. ELLIE A. HILL.

Sugar Cookies.

One and two-thirds cups sugar, one cup butter, three eggs, two-thirds teaspoonful soda, one small nutmeg, flour to roll. Roll thin and bake *in a quick oven.*

Sugar Cookies.

Two eggs, one cup sugar, two thirds cup of butter and lard, one teaspoonful cream-tartar, one scant teaspoonful soda, two tablespoonfuls cold water. Flavor with lemon; flour to roll. Roll thin. Bake in quick oven.

MRS. M. A. EVERETT.

Cornstarch Cake.

Two cupfuls of sugar, two cupfuls of sifted flour, one cupful of cornstarch sifted in flour, one cupful of butter, beaten to a cream with sugar, one cupful sweet milk, two even teaspoonfuls baking powder sifted in flour, and the whites of eight eggs beaten to a stiff froth and added the last thing. Flavor to the taste. Bake in a slow oven three-quarters of an hour. MRS. M. L. T. HIDDEN.

Coraline Cake.

Half a cupful of sweet milk, half a cupful of rich cream, one cupful of sugar, one egg, two cupfuls of graham flour, one teaspoonful of baking powder. Bake in two tins. When done split open with a sharp knife and fill in with raspberry or strawberry juice that has been thickened with cornstarch or gelatine. By using boiled custard for filling it will make what cooks call French pie.

ALICE B. STOCKHAM, M. D.

Cream Cake.

One cupful butter, two cupfuls powdered sugar, one cupful sweet milk, three cupfuls flour, one teaspoonful soda, two teaspoonfuls cream of tartar, whites of four eggs beaten to a stiff froth. Flavor with orange or vanilla.

MRS. E. R. ABBOT.

Cream Cake.

One cup sweet cream, two cups sugar, three cups flour, four eggs, one teaspoonful soda, two teaspoonfuls cream of tartar. Bake in a moderately heated oven. It is better baked in a *shallow* pan. One-half the above makes an excellent "Washington pie." This cake steamed and liquid pudding-sauce added makes an excellent pudding. The cake, in a stone jar, will keep several weeks without drying.

L. Clementine Gates.

French Cream Cake.

One cup sugar, three eggs, one and a-half cups flour, two tablespoonfuls cold water, one teaspoonful cream of tarter, half a teaspoonful soda. Make three layers in Washington-pie plates.

Cream.—One egg, half a pint of milk, three heaping tablespoonfuls sugar, two even tablespoonfuls flour and a little salt. Flavor with lemon or vanilla. Stir into the milk when hot the flour mixed in a little cold milk, also the beaten egg and sugar. The same cake may be used for jelly, and for chocolate Washington pie, only add butter the size of an egg.

Chocolate Filling.---Half a cake of chocolate, grated, one cup sugar, one egg, one tablespoonful cold water. Stir all together, and steam ten minutes, as for boiled custard. Put on the cake while warm.

Mrs. Alice M. Southwick.

Plain Cream Cake.

One-half cupful cream, two-thirds cupful of sugar, one egg beaten to a froth, two cupfuls of flour, one teaspoonful soda, one teaspoonful cream of tartar. Bake in little cups twenty minutes or in a pan forty minutes.

Mrs. W. D. Forbes.

Cup Cake.

Two eggs, one cupful sugar, one-half cupful butter, two-thirds cupful milk, two and one-half cupfuls flour, one-half teaspoonful saleratus, one teaspoonful cream of tartar, one-half good-sized nutmeg. Beat the butter and sugar to a cream, and add the eggs, well beaten; dissolve the saleratus in the milk. Beat the whole together and bake in a moderate oven three-fourths of an hour. By adding berries this makes an excellent berry cake, or it may be baked in hearts and rounds, putting a few currants on the top of each cake. L. F. S. BARNARD.

Delicate Cake.

Two cupfuls of sugar, three-fourth cupful of butter, one cupful of milk, three cupfuls of flour, two level teaspoonfuls of baking powder, the whites of six eggs. Sift flour and baking powder together. EMILY S. BOUTON.

Delicate Tea Cake.

One and one-half cupfuls sugar, one-half cupful butter, the whites of three eggs beaten to a froth, half a cupful of milk, three cupfuls flour, one teaspoonful cream of tartar, one-half teaspoonful soda. Flavor with lemon or vanilla.

MRS. S. J. VINCELETTE.

Dough Cake.

Two cupfuls raised dough ready for baking; one cupful sugar and two-thirds cupful butter, beat well and stir into the dough. Add a little nutmeg and cinnamon, one-half teaspoonful soda, two well beaten eggs, and one cupful of raisins rubbed in flour before mixing in the dough. This is the last to be added with as little mixing as possible. Put into the oven without rising again.

MRS. D. W. FORBES.

Drop Cakes.

One-half cupful butter and two cupfuls sugar beaten together, three eggs, one cupful milk, one teaspoonful soda, and two of cream of tartar rather scant, two and one-half cupfuls of flour and a dessert spoonful of caraway seeds. Drop a teaspoonful in the pans for each cake; bake in a hot oven. MRS. ZILPHA H. SPOONER.

Eggless Cake.

One cupful sour milk, one teaspoonful soda, one cupful sugar, one-half cupful butter, one cupful raisins chopped, cinnamon and nutmeg, flour to stiffen.

MISS M. L. MORELAND.

Mother's Election Cake.

Five pounds flour, six eggs, two pounds sugar, one pint yeast, three-fourths pound butter, one quart sweet milk, three-fourths pound lard, six nutmegs. Take about three pounds of the flour, and about one-third of the sugar, and stir up with the yeast and two-thirds of the milk, to rise over night, or until it begins to fall on the top; then add the rest of the ingredients and bake in loaves about the same as you would bread. MISS M. A. HILL.

English Lunch Cake.

Cream one and a half cupfuls of butter and the same of sugar with three eggs; add one and a half cupfuls of milk, four cupfuls of flour, two cupfuls of currants, citron if liked, half a teaspoonful of mixed spices and a teaspoonful of soda. If baked in one loaf, it will take about an hour—in two pans, about half an hour. MISS E. B. PLYMPTON.

Feather Cake.

One cupful sugar, one-half cupful butter, one-half cupful cornstarch, one cupful flour, whites of three eggs, one teaspoonful lemon essence, two teaspoonfuls baking powder (or one teaspoonful cream of tartar and one-half teaspoonful soda), one-half cupful sweet milk.

HARRIETT C. BATCHELDER.

Feather Cake.

Beat to a cream half a cup of butter, add to this two cups of sugar, and beat well together; add one cup milk with one teaspoonful soda dissolved in it; beat well. Then add one cup sifted flour with two teaspoonfuls cream of tartar previously rubbed into it. Add next the well-beaten yolks of three eggs. Beat the whites separately until stiff; add them, and then two more cups of flour. Beat well between each addition. MRS. MARY F. CROWELL.

French Cake.

Three cups sugar, one cup butter, beaten together, four eggs, beaten separately, and added, one and one-half cups of milk, five cups flour, one teaspoonful soda, two of cream-tartar, flavor to taste; makes two loaves.

MRS. ALICE M. SOUTHWICK.

Frosting.

One cup sugar, one-third cup milk; boil about four minutes, or until it drops like syrup from the spoon. Take from the fire and beat until it stiffens, and is of the right consistency to pour and spread quickly over the cake.

MRS. S. A. RICHARDS.

Boiled Frosting for Cake.

Two-thirds of a cup of sugar to the white of one egg. Cover the sugar with cold water in a sauce-pan, boil, not stirring, until, when tried in a little cold water, it will break like candy. Beat the white of the egg very light, pour the boiling syrup into it slowly, beating hard all the time. Continue to beat until stiff enough to put on the cake.

MRS. M. M. WOOLFORD.

Fruit Cake.

Two cups sugar, one cup molasses, two cups butter, six eggs, five cups flour, one teaspoonful soda, one teaspoonful of each kind of spice, one pound citron, one and one-half pounds raisins, one pound currants. Make two loaves; bake four hours in a slow oven.

MRS. ZILPHA H. SPOONER.

Fruit Cake.

Stone and chop one pound of raisins, one pound of currants, and slice half a pound of citron. Beat five eggs and add two cups of brown sugar, beat well; add one cup of butter, one of molasses, one-half cup of sour milk, one large teaspoonful of soda, one teaspoonful of all kinds of spices, six cups of flour, reserving one-half cup of the flour to mix with the fruit to prevent it from settling; add the fruit last. Place buttered letter-paper in the bottom of the tins.

MRS. M. E. SAMMET.

Fruit Cake

One-half cupful each of brown sugar, butter, molasses, and milk, one egg, two cupfuls of flour, one cupful of currants, one cupful of raisins, one-quarter pound of citron, one-half teaspoonful of soda, one-half teaspoonful each of nutmegs, cinnamon and cloves.

MRS. B. J. STONE.

Fruit Cake.

One and one-half cupfuls butter, two cupfuls brown sugar, one and one-half cupfuls molasses, one cupful sweet milk, four eggs, one teaspoonful soda, one tablespoonful cloves, one tablespoonful allspice, one tablespoonful cinnamon, one nutmeg, one-half pound citron, two pounds raisins, one pound currants. Flour to make it sufficiently stiff to hold up the fruit, (about six cupfuls of flour). Bake in an oven not very hot.

FROSTING.—Whites of three eggs, two even teaspoonfuls of powdered starch, one pound powdered sugar, extract to suit taste. AMANDA M. LOUGEE.

Lowell Fruit Cake.

Five coffee cups of flour, five eggs, two cups of molasses, one cup brown sugar, one cup butter, one teaspoonful soda, one pound of raisins, a half pound of currants, and a quartea pound of citron if liked. MISS E. B. PLYMPTON.

Plain Fruit Cake.

One cup of butter, one cup sweet milk, two cups brown sugar, three cups flour, four eggs, one pound raisins, one-half pound currants, one-fourth pound citron, three teaspoonfuls baking powder, cinnamon and nutmeg to taste. I sometimes substitute one-fourth pound figs for the one-half pound currants. This cake is good in three days, will keep three months. LOUISE V. BOYD.

Gingerbread.

Two-thirds cup butter, one cup sugar, one cup milk, one cup molasses, two eggs, four cups flour, one teaspoonful ginger, one teaspoonful soda in the molasses, two teaspoonfuls cream of tartar in the flour. EMILY A. FIFIELD.

Ginger Snaps.

Put into a large cup one teaspoonful of soda, two tablespoonfuls hot water and three tablespoonfuls melted butter. Fill the cup with molasses, add ginger; mix hard, roll thin and bake in a quick oven.

MRS. S. C. WRIGHTINGTON.

Molasses Gingerbread.

One cup molasses, one teaspoonful soda in the molasses, one tablespoonful melted butter one-half cup cold water, small teaspoonful ginger, flour for soft batter.

MRS. M. A. EVERETT.

Molasses Gingerbread.

One cup molasses, one even teaspoonful soda, one cup hot water, one-half cup of butter, flour to make it the consistency of cake if to be baked in small tins; if to be rolled add a little flour. A pinch of ginger, cassia and clove, or if ginger alone, a tablespoonful. Dissolve the soda in a spoonful of water and stir it in the molasses till it foams, then add the other ingredients. MRS. E. C. CROSBY

Short Gingerbread.

One cup of milk, two cups sugar, one of butter, one-half teaspoonful of saleratus, one tablespoonful of yellow ginger, flour enough to roll. Roll very thin.

MISS C. WELLINGTON.

Soft Gingerbread.

One cup molasses, piece of butter the size of an egg, two-thirds cup of milk, two cups flour, one teaspoonful soda dissolved in one-third cup of boiling water.

MRS. FORREST W. FORBES.

Ginger Snaps.

One cup butter, one cup sugar, one cup molasses, two-thirds cup water, one heaping teaspoonful soda, one teaspoonful ginger, and a plenty of flour. Roll very thin. Mix butter and sugar together, add molasses and then the water, reserving a little in which to dissolve the soda, which should be put in after a part of the flour has been used.

MRS. WILLIAM C. COLLAR.

Ginger Snaps.

One cup molasses, one cup butter and one-half cup sugar, boiled together, then add as little flour as possible to roll out, one-half teaspoonful soda, two teaspoonfuls ginger.

S. LOUISA SIMONDS.

Graham Crackers.

Take one part cream to four parts milk, mix with flour, as soft as can be handled, knead twenty minutes, roll very thin; cut square or round and bake quickly twenty minutes. Handle carefully while hot, pack away when cool in a stone jar.

ALICE B. STOCKHAM, M. D.

Graham Wafers.

Use Akron flour, mix with pure cold water, no salt, knead thoroughly fifteen minutes, roll very thin, about half as thick as soda crackers, cut in two-inch squares and bake quickly. These will keep for months in a dry place. It makes them crisp to place them in the oven a few minutes before bringing them to the table.

ALICE B. STOCKHAM, M. D.

Hedge Cake.

Break two eggs into a cup, fill it up with cream, then beat light, add one cup of sugar, one and one half cups of flour, one-half teaspoonful soda, one teaspoonful cream of tartar; bake in three jelly pans.

CREAM.—One-half pint of milk, boil it and add two eggs and the yolk of another beaten with one-half cup of sugar and one large spoonful cornstarch or flour; flavor.

FROSTING.—The white of one egg and one cup fine sugar.

MRS. RICHARDSON.

Hermit Cakes.

One and one-half cups butter, two cups sugar, three eggs, one teaspoonful each of clove nutmeg and cinnamon, one cup stoned raisins or currants, four tablespoonfuls milk with one teaspoonful soda, flour enough to roll out (not too stiff) a third of an inch thick. Cut in squares and bake in a quick oven. MRS. WILLIAM C. COLLAR.

Hermit Cakes.

Two eggs, one and one-half cups sugar, three-fourths cup butter, one cup currants, two tablespoonfuls sweet milk, one teaspoonful cloves, one teaspoonful cinnamon, one teaspoonful nutmeg, flour enough to roll thin and cut like cookies. MISS H. B. HICKS.

Jelly Roll.

One cupful sifted flour, two-thirds cupful sugar, one teaspoonful baking powder, one-half teaspoonful salt, two eggs, two large spoonfuls rich milk; beat thoroughly together, and bake to a light brown. Spread while warm with any kind of jelly, and roll closely in clean white cloth. This may be used for cream cake, or frosted with chocolate; either will be *good.*

MRS. J. BLACKMER.

Loaf Cake.

Half a cupful sweet cream, one cupful butter, two and a quarter cupfuls sugar, three and a half cupfuls flour, one cupful raisins, five eggs and one teaspoonful soda; mix butter and sugar thoroughly, add cream and soda (no cream of tartar used); then add the eggs well beaten, flour, raisins well floured, and spices to suit the taste; bake in shallow pans in a *slow* oven. This quantity will make two loaves. It will keep good for six months, and is improved by age.

PAUL GATES.

Magic Cake.

One-half cupful butter, one cupful sugar, one and a half cupfuls flour, three eggs, three tablespoonfuls of milk, one teaspoonful of cream of tartar, one-half teaspoonful of saleratus; flavor with almond essence.

MISS L. F. S. BARNARD.

Marble Cake.

FOR THE DARK.—Yolks of seven eggs, one cup molasses, one teaspoonful cloves, five cups flour, one teaspoonful allspice, two cups brown sugar, one teaspoonful soda, one cup sour cream, two teaspoonfuls cinnamon, one cup butter, and one nutmeg.

FOR THE LIGHT—Whites of seven eggs, one cup of butter, two of white sugar, three of flour, one-half cup of sweet milk, one teaspoonful cream of tartar, half a teaspoonful of soda. Always cream butter and sugar together and beat eggs separately.

MRS. ALICE M. SOUTHWICK.

Marble Cake.

LIGHT PART.—Whites of three eggs, one-half cup of butter, one cup of sugar, one-third cup of sweet milk, two cups of flour, one-third teaspoonful of soda, two-thirds teaspoonful of cream of tartar, nutmeg and salt.

DARK PART.—Yolks of three eggs, one-third cup of butter, one cup of molasses, two cups of flour, one teaspoonful of soda; add tablespoonful of ginger, plenty of fruit, spice and some salt. Drop into the pan first a spoonful of the light part, and then one of the dark, etc. Makes good-sized loaf.

MISS L. F. S. BARNARD.

Marble Cake.

A nice marble-cake may be made from recipe for chocolate cake on page 79, by using only half the amount of chocolate, stirring it into half the cake mixture. Then proceed as is usual with marble-cake.

MRS. E. E. KELSEY.

New Jersey Molasses Cake.

Into the mixing-bowl, pour two cups molasses, two thirds cup lard, lard and butter mixed, or sweet pork drippings; *dissolve* teaspoonful saleratus or baking-soda in one cup cold water. Stir all well together; add enough sifted flour to make a moderately stiff batter. Care must be taken not to mistake the foam from soda for a consistency caused by flour. Bake *immediately* in two well-buttered pans in a not too-hot oven. Any kind of spice may be added; also raisins if liked. When carefully made it is delicious. MRS. H. O. HAWKINS.

Mother's Cake.

One cupful sugar, half cupful butter, half cupful milk, two cupfuls flour, three eggs; cream the butter and sugar; beat the yolks and whites of the eggs separately, then add all the ingredients, beating them well, and add two teaspoonfuls baking powder. Bake in a round, scalloped pan, and when done sprinkle powdered sugar over the top.

MISS E. A. RANSOM.

Nut Cake.

One cupful butter, two of sugar, one of milk (scant), six eggs (beat whites separately, and save white of one for frosting), one teaspoonful of cream-tartar, one-half of soda, three and one-half cups of flour, one cupful meats (takes one pound nuts), one-half teaspoonful of lemon, one-half teaspoonful of vanilla. This makes two loaves.

ANNA B. TAYLOR, M. D.

Park Street Cake.

One-half cup butter, one cup milk, two cups sugar, three cups pastry flour, four eggs, one-half teaspoonful soda, one teaspoonful cream of tartar, one teaspoonful vanilla, little mace. Rub the butter in a warm bowl until like cream, and add one cup of the sugar gradually, add the remaining cup of sugar to the beaten yolks of the eggs, beat until very light, and add to the butter; add the flavoring; then beat the whites stiff and dry, and let them stand while you add a little milk and flour alternately to the mixture. Add the whites last. This makes two loaves.

MRS. E. R. ABBOT.

Plum Cake.

Three-fourths pound sugar, three-fourths pound butter, six eggs, one pound flour, one pound raisins, one-fourth pound citron, one wine-glassful of raisin water, pinch of soda. Boil raisins and citron together and let the water evaporate until only the wine glassful is left. Stone the raisins. A little powdered clove or nutmeg to taste. Bake about an hour. This quantity makes two loaves.

MRS. SARAH S. SWAIN.

Polka Cake.

Two cups of sugar, one cup of butter, one cup of molasses, one cup of milk, four eggs, one teaspoonful of soda, chopped raisins and spices. EMILY A. FIFIELD.

Pound Cake.

One pound sugar, one-half pound butter, three-quarters pound flour, eight eggs. Beat the eggs separately, cream sugar and butter together, then stir in the yolks, then the whites, being well beaten, beat all together a few minutes, then *stir* in the flour, but do not beat it. Bake in a moderate oven twenty minutes. This receipt makes twenty-eight rounds. KATHARINE STARBUCK.

Irish' Wedding Cake.

One and one-half cups sugar and one cup butter, beat to a cream; add the beaten yolks of two eggs, three-fourths of a cup of milk less two teaspoonfuls reserved for dissolving soda, two and one-half cups of flour, half a nutmeg, half pound raisins, quarter pound currants, half teaspoonful soda; lastly add the beaten whites of two eggs. Stir well.

MRS. H. ANDREWS.

Queen's Cake.

Six eggs, six cups flour, four cups sugar, one cup butter, two cups cream, two teaspoonfuls soda. Use lemon or spice and fruit. MRS. SARA T. L. ROBINSON.

Raisin Cake.

One egg, one cup sugar, one-half cup butter, one cup sour milk, one cup raisins, chopped, one pint of flour, one teaspoonful saleratus. MRS. F. D. OSGOOD.

Ribbon Cake.

Two cups sugar, three eggs, two-thirds cup butter, one cup sweet milk, three cups flour and one heaping teaspoonful baking powder put into the flour and sifted. Add a little salt and flavor with lemon or almond. Put half the above into two oblong pans. To the remaining add one teaspoonful molasses, one large cup raisins chopped and stoned, one-quarter pound citron sliced, one teaspoonful cinnamon, one-half teaspoonful each of cloves and allspice, a little nutmeg, one teaspoonful flour; put into two pans. Put the sheets together, while warm, alternately with a little jelly between; cut into thin slices for the table, will cut most easily the day it is baked. It may be baked in one large pan without the fruit, pouring in the dark and light in alternate layers. MRS. E. R. ABBOT.

Smith Cake.

One cupful of butter, two of milk, three of sugar, four eggs, five and one-half cups flour, one cup currants, one teaspoonful lemon, three teaspoonfuls yeast powder. Cream the butter, add half the sugar, and the yolks of the eggs, well beaten; add the remainder of the sugar, and the milk, and flour (with powder sifted in it) alternately; then add the whites of the eggs, beaten stiff, and lastly the currants, rolled in the half cup of flour. Fill pans with the mixture an inch and a-half thick, and bake in a quick oven from thirty to forty minutes. JUDITH WINSOR SMITH.

Spice Cakes.

One egg, two-thirds cupful molasses, two-thirds cupful sugar, two-thirds cupful melted butter, one cupful milk, two and one-half cupfuls flour, half teaspoonful soda, one teaspoonful cream-tartar, one teaspoonful mixed spice, (clove, cinnamon, nutmeg,) one tablespoonful vinegar. Bake in muffin rings or gem pans.

MRS. MARTHA J. WAITE.

Spice Cake.

Two eggs, one and one-half cupfuls sugar, half cupful butter, half cupful milk, two and one-half cupfuls flour, half cupful raisins, a little of all kinds of spice, quarter teaspoonful soda. HARRIET C. BATCHELDER.

Sponge Cake.

Weigh your eggs (five being enough for an ordinary cake-tin); weigh the shells; deduct their weight from the weight of the eggs; add the remaining weight of sugar, and half this weight of flour. Beat the yolks and sugar well together, then beat the whites to a stiff froth and stir in the sugar and yolks thoroughly, then stir in gently the flour and flavoring, and bake in a moderate oven. No baking powder is necessary. MRS. EMMA E. FOSTER.

Sponge Cake.

Weigh six unbroken eggs, separate the yolks and beat them ten minutes, add the weight of the unbroken eggs of sugar; beat the whites until light, add them, with the weight of three unbroken eggs of flour, the juice of one-half of a lemon, the rind of a whole lemon grated.

MRS. M. E. SAMMET.

Sponge Cake.

Four eggs, two small cups of sugar, two cups of flour. Eggs and sugar well beaten together; stir the flour in as quickly as possible; bake twenty minutes in a quick oven.

MARY J. WILLIS.

Sponge Cake.

One pound pulverized sugar, one-half pound flour, twelve eggs, one lemon, grated rind and juice. Beat whites to a stiff froth, beat also the yolks, then beat thoroughly together, adding sugar. Grate lemon rind into flour and stir in lightly, but do not beat the mixture after flour is added. If not convenient to put in lemon, add half a teaspoonful baking powder to the flour and put in the extract of lemon. Bake in a moderate oven.

MRS. THOS. S. LYON.

Sponge Cake.

Four eggs, one cup of sugar, one cup flour sifted four times. Beat the yolks and sugar together, then add the flour, and lastly the whites beaten to a stiff froth. Bake in a slow oven. MRS. MARCIA E. P. HUNT.

Berwick Sponge Cake.

Beat three eggs five minutes, add one and one-half cups of sugar; beat five minutes; one cup of flour with one teaspoonful of cream-tartar stirred dry in the flour; beat one minute; half cup of cold water, with half teaspoonful of saleratus well dissolved in it; beat one minute; the rind of a lemon, juice of half a lemon; beat one minute. Add another cup of flour, beaten in as light and quick as possible.

MRS. JOSIE CURRIER.

Eureka Sponge Cake.

Four eggs beaten, with one and one-half cups of sugar, two cupfuls of sifted flour, baking powder and lemon extract, each one teaspoonful. Beat thoroughly together, and add three-fourths cupful of boiling water. Is very thin, but makes a delicious and wholesome cake. Is good made from white or graham flour. Makes a nice layer cake by baking it in jelly tins. CORA L. STOCKHAM.

Mrs. Holt's Sponge Cake.

One coffee-cupful of sugar, or about one and one-quarter teacupfuls, one coffee-cupful flour, sifted before using, five eggs, yolks and whites beaten separately, one tablespoonful cold water, or lemon-juice. Stir in flour as gently and carefully as possible, and do not beat thereafter.

MRS. H. O. HAWKINS.

Rocky Mountain Sponge Cake.

Two cups of sugar, yolks of five eggs, whites of three, two cups of flour, one-half cup of hot water, one teaspoonful baking powder. Beat the eggs well, then add sugar and beat together until light, add the flour and baking powder, stir gently. When well mixed add the hot water. Bake in a quick oven. MRS. M. M. WOOLFORD.

Superior Sponge Cake.

Beat four eggs very light and mix them together, stir in two cups of pulverized sugar, the juice and half of the grated rind of a lemon and a pinch of salt, add one cup of flour, stirring gently, then another cup in which mix two *even* teaspoonfuls of baking powder; lastly, stir in, a very little at a time, three-fourths of a cup of almost boiling water; bake in deep, narrow pans. If iced, this makes as good sponge cake as any one could desire.

CATHERINE H. BIRNEY.

Swiss Cake.

One and one-half cups sugar, one cup milk, three cups flour, four tablespoonfuls butter, two eggs, one and a half teaspoonfuls cream of tartar, one teaspoonful soda.

MRS. B. M. NICHOLS.

Vanilla Crisps.

One cup of butter, one cup of sugar, one-half cup of milk, one egg, one-half teaspoonful of vanilla, one-half teaspoonful of soda, six cups of flour; roll very thin.

MRS. D. P. WASHBURN.

Vanilla Wafers.

One cupful sugar, two-thirds of a cupful of butter, four tablespoonfuls of milk, one of vanilla, one egg, one and a half teaspoonfuls of cream of tartar, two-thirds of a teaspoonful of soda, flour enough to roll out very thin.

SARA A. LOUD.

Vanilla Wafers.

One cupful sugar, two-thirds cupful butter, four table spoonfuls milk, one tablespoonful vanilla, one egg, one teaspoonful cream-tartar, one-half teaspoonful of saleratus or soda, and flour to roll very thin.

MRS. B. M. NICHOLS.

Walnut Cake.

One cup sugar, one-half cup butter, one and three-fourths cupfuls flour, two eggs, one-half teaspoonful soda, one teaspoonful cream-tartar, and one cup walnuts chopped not very fine. MRS. FORREST W. FORBES.

English Walnut Cake.

Four eggs, two cupfuls of sugar, one cupful of butter, one cupful of milk, three cupfuls of flour, one teaspoonful of cream-tartar, half teaspoonful of saleratus, one pound English walnuts. Mix the butter and sugar to a foam with the hand; add the walnuts as you would citron, in pieces. Makes two small loaves. MRS. JOSIE CURRIER.

Washington Cake. St. Louis, 1780.

Two cups butter, three cups sugar, four cups flour, two teaspoonfuls baking powder, five eggs, one cup milk, one cup stoned raisins, one-half cup washed and picked currants, one-fourth cup chopped citron, one teaspoonful each of royal extract of nutmeg and cinnamon. Beat the butter and sugar to a white, light cream, add the beaten eggs gradually, the flour sifted with the powder, milk, raisins, currants, citron, and extracts. Mix into a smooth, medium batter, and bake in a shallow square cake pan in a rather quick, steady oven, one and one-half hours; when cold, ice with white icing.

White Icing.—The whites of four eggs, one and one-half pounds of white sugar dust, one-half teaspoonful acetic acid (or the juice of half a lemon), one-half ounce of royal extract of rose. Place the whites with the sugar in a bowl, with the acetic acid and extracts. Beat with a wooden spoon until, letting some run from the spoon, it maintains the thread-like appearance several minutes, when use as directed.

Mrs. S. J. Vincelette.

Wedding Cake.

Four eggs—frost the top and sides—one and one-half cupfuls of butter, two cupfuls of sugar, one cupful of molasses, one-half cupful of milk, four cupfuls of flour, one teaspoonful saleratus, four teaspoonfuls cloves, four teaspoonfuls cinnamon, five or six nutmegs, one pound currants, one and one-half pounds raisins stoned and chopped, one-half pound citron. Before using the flour brown it in the oven to make the cake dark. Work the butter and sugar to a foam with the hand. Mix the fruit with part of the flour. Frost the top and sides.

Mrs. F. D. Osgood.

Welcome Cake.

Beat together one and one-half cupfuls of sugar and one-half cupful of butter; add three well beaten eggs and three small cupfuls of flour in which a teaspoonful cream of tartar and a half-teaspoonful soda have been sifted; beat in a half-cupful of milk and a cupful of currants or seeded raisins.

MISS E. B. PLYMPTON.

White Cake.

One cupful of butter, one and two-thirds cupfuls of sugar, whites of six eggs, two teaspoonfuls cream of tartar, one teaspoonful soda, one cupful of sweet milk, flavor with lemon extract.

For Fruit Cake, made by taking a part of the above mixture, take about one-half, and add one cupful of seeded raisins, and a small handful of the same boiled, cinnamon and nutmeg for flavoring according to taste.

MRS. H. O. HAWKINS

White Mountain Cake.

One and a-half cups sugar, and two tablespoonfuls butter beaten to a cream, yolks of two eggs, one cup milk, three cups flour, two teaspoonfuls cream of tartar, one teaspoonful lemon, half teaspoonful soda dissolved in milk; add beaten whites of two eggs. Bake in a quick oven.

MRS. H. ANDREWS.

White Mountain Cake.

Whites of two eggs, one and one-half cupfuls sugar, one cupful milk, half cupful butter, two teaspoonfuls of baking powder, three cupfuls of flour. Bake in jelly tins. For frosting for layers use whites of two eggs (beaten light), one-half pound of pulverized sugar; sprinkle cocoanut over each layer after the frosting has been used.

MISS M. L. MORELAND.

Belle's White Mountain Cake.

Whites of six eggs, well beaten, one and one-quarter cupfuls sugar, one and one-quarter cupfuls flour, half cupful butter, half cupful sweet milk, half cupful cornstarch, one teaspoonful soda, two teaspoonfuls cream-tartar. Bake as jelly cake, with either cocoanut or frosting between layers.

MRS. H. O. HAWKINS.

PRESERVED AND CANNED FRUITS, JELLIES, ETC.

Fruit Canning Made Easy.

Let the jars be washed out and thoroughly dried after the last season's use, and then put away each with its own cover screwed on; thus they are usually sweet and clean and the trouble and delay at time of canning which would arise from mixing of covers or from musty or dusty jars is avoided. Fold a towel smoothly into several thicknesses and wet it thoroughly in hot or cold water. Lay it in a shallow tin and set the jar upon it, taking care that it rests smoothly upon the towel, which must not touch the sides of the jar. Close windows and doors so as to avoid drafts. A careful observance of these directions will prevent the jars from breaking. Set the tin containing the jar on the back of the stove close to the kettle of boiling fruit, dip the fruit into the jar while boiling, fill just to the brim, wipe off any juice that may have run over on the rubber or screw thread and seal at once, screwing tightly. It is well to turn the jar upside down at first, to be sure no juice could escape, as sometimes there is fault in cover or rubber, and the jar, of course, would not be air-tight if not water-tight. The dread of fruit canning is to a large extent removed by the practice of this simple process, which is sure to be successful if the directions are carefully followed.

MRS. E. E. KELSEY.

To Can Fruit in Wide-Mouthed Bottles.

Prepare your fruit as you would to fill glass jars. Then take any wide-mouthed bottle (or a jar which may have lost its cover), heat it, and then fill it full with boiling fruit. After wiping the bottle round the top, take a piece of either brown or white soft paper, or soft white cloth, and tie it over the top with fine cord. Then take a piece of cotton-batting (not wadding) and tie that on in the same way, over the cloth or paper. Afterward tie a second piece of paper over the batting so that the latter may not get pulled off. Paste on the bottle a label giving kind of fruit and date of putting up. Fruit put up in this way will keep as well as if put in Mason jars, and has the merit of economy, as housekeepers often have wide-mouthed bottles which they could use for this purpose. I have fruit, prepared in this way, which is as good now as when put up two years ago. MRS. O. A. CHENEY.

To Can Baked Apples.

Pare, quarter, and core the apples, and bake tender and brown in a moderate oven; boil the parings, or use *tart*, defective apples you may have on hand, to make a sufficient syrup to cover the fruit in the cans; strain the parings or apples, adding sugar to the juice, and boil till it is almost ready to jelly; seal while hot, and the dish will be delicious in the springtime when your winter apples are all gone. I use glass jars. LOUISE V. BOYD.

To Can Barberries.

Pick off all the stems, or not, as you please, put two quarts water into a porcelain kettle, add two quarts of the barberries; no sugar, it will toughen them; boil five minutes, skim out into an earthen jar or dish; now put in more berries, boil five minutes, skim out, and keep on in the same way until you have boiled all your berries; then add one teacupful of granulated sugar to every quart of berries, return to the kettle and let the whole mass cook ten minutes, stirring carefully all the time to evenly mix the sugar and get them all hot. Fill the glass jars running-over full, being careful that the rubbers are all right, and pressed tightly down. Screw on the covers tight, and when cool, tighten again. This will be very nice for the sick. More sugar can be added when used if desirable, but in fevers and at many other times when the appetite is poor, they are better not too sweet. DR. ALICE M. EATON.

To Cook Cranberries.

For every quart cranberries take one pint water and one pint sugar. Have the water in your preserve kettle (porcelain lined) *boil*. Add the sugar, stir well and *boil fast* one minute; then put in the berries and let them stand in the syrup hot as possible, *without boiling*, about five minutes, stirring them a little; then let them have the heat and stew *rapidly* for fully five minutes more. Take from the fire, and let them stand hot for at least half an hour before pouring off the kettle. MRS. MARY S. TARBELL.

Apple Jelly.

Windfall apples of various kinds may be used together. Cut the apples into quarters, take off their stalks and blossoms, but do not peel them. Fill a large preserving pan with the apples thus prepared, add a small teacupful of water and the peel and juice of a lemon; stew till the fruit is quite soft and of the consistency of porridge. Have ready a bag, made of a square of flannel, in the shape of an extinguisher, with four strings sewn to the hem at the open end. Tie this bag to the four legs of a small table, turned top downwards on the floor, allowing room beneath the point of the bag for a basin to receive the droppings from the apples. Pour the hot mash into the jelly bag and leave it in a warm place for twelve to twenty hours; the bag must not be squeezed to press out the juice. To one pint of juice add one pound of loaf sugar and boil about half an hour, till it jellies, when tried on a cold plate. Pour into glasses, or cups, or shapes; the jelly will turn out quite firm and will keep for years if covered from the air. Several pans full of apples will be needed for a boiling of jelly.

REBECCA MOORE.

Barberry and Apple Jelly.

Apples cleaned of specks and stalks and cut into a porcelain kettle to be simmered in a little water till soft like apple sauce. To press out the juice it is easier to put a small portion at a time into the jelly bag. Have also a large earthen pitcher full of washed barberries set in a pot of boiling water; boil till the juice begins to run out; strain these through jelly bag. About half apple and half berry juice is the best proportion; measure before boiling. Boil half an hour, skimming meanwhile; add equal measure sugar, boil ten minutes, or till a few drops will stay in a spoon turned upside down. Excellent with meats.

MISS H. B. HICKS.

Grape Jelly.

Pick off the grapes into a pail, cover, set them into a kettle of water and cook until the juice runs; then pour them into a strainer cloth, tie up, and hang up on a nail to drain; let them drain all night; then measure out the same amount of granulated sugar that you have of juice, put into a porcelain kettle and boil hard twenty minutes; try a little on a cold dish, and if it does not jelly boil it longer and try again.

MRS. ALICE M. SOUTHWICK.

Orange Jelly.

INGREDIENTS.— Nelson's gelatine, four oranges, two lemons, one cupful of sugar, water. Dissolve one-half box of Nelson's gelatine in half-pint of tepid water; then add a pint of boiling water, one cupful of sugar, the juice of four oranges and two lemons; strain this mixture and pour into a mould or into jelly tumblers. This quantity will make four ordinary jelly tumblers.

MRS. HARRIET M. TURNER.

Delicious Grape Jelly.

Put the fruit in a preserving kettle and set it on the stove where the fruit will heat gradually; after the juice is extracted, strain through the jelly bag; do not press the fruit but let it drip through thoroughly; put one pound granulated sugar to every pint of clear juice; heat the sugar hot and dry in the oven; boil the juice five minutes, then add the hot sugar; let it come to boiling point, and then remove from the stove; pour in jelly glasses, seal tightly when cold, and keep in a dry, cool, dark place. The grapes remaining in the jelly bag can be pressed, and the juice extracted will make a second grade of jelly. MRS. S. C. VOGLE.

Transparent Jellies. (No. 1.)

To make transparent jellies from strawberries, raspberries, currants, grapes, blackberries, or any small fruit, carefully select that which is sound and not too ripe, and without adding water, put to press over night, previously bruising in a wooden bowl with your potato masher, and using as a strainer a jelly-bag make of cheese cloth. You will find your clothes wringer a most convenient press for extracting the juice, as it will take the open end of the bag between the rollers and hold it securely at any stage of pressure desired. Allow the juice to escape into earthen or wooden vessels, but on no account into tin or other metal, as the contact would color the jelly, and it is your object to make it clear. The next morning measure the juice, allowing a cupful of granulated sugar to each cupful of the liquid if the fruit is sour, or three cupfuls of sugar to four of juice if the fruit is sweet, and boil rapidly in a porcelain kettle for five minutes. Dip carefully into glasses, and if not thoroughly stiff when cold, set away for a few days in the sunshine.

This recipe will never fail if the directions are strictly observed. ABIGAIL SCOTT DUNIWAY.

Transparent Jelly. (No. 2.)

To make transparent jelly from apples, peaches, pears, quinces and crabapples, cut the sound, barely ripe fruit into quarters without paring (though it is well to remove the seeds), and place in a porcelain kettle. Add boiling water to cover the fruit and cook rapidly till soft, adding more boiling water if necessary. Then place in a new jelly-bag and hang up to drip into earthen vessels. The juice, which for such fruits should be colorless, is to be treated as for making jellies of small fruits, except that it should be boiled after sugar is added till it turns to a bright amber—perhaps twenty minutes.

ABIGAIL SCOTT DUNIWAY.

Marmalade.

To make a delicious marmalade, squeeze the pulp soft from any of the above processes through the thin meshes of the jelly-bag and add sugar in same proportion as for jelly. Flavor if desired with the juice of lemons with the rinds added, chopped fine; or put a bag of different kinds of whole spices in the boiling mass, removing when the marmalade is done. Let it cook an hour.

ABIGAIL SCOTT DUNIWAY.

Lemon Butter.

One lemon, one cup sugar, three eggs; beat all together. Nice for tea. MRS. S. W. FULLER.

Plums to Eat with Meats.

Half a pound of sugar to one pound of common small blue plums. Place plums and sugar alternately in a two-quart stone jar until full, plums at the bottom, a layer of sugar on top; set the jar in an oven of bread heat, leaving the door open; bake for several days, keeping watch they do not burn; when of a preserve-like consistency, take from the oven, cover them well and keep in your sweetmeat closet. Plums thus prepared have a peculiar flavor of their own, and are especially nice with roasts. It is best to start about three jars at once, removing all to one jar as they bake away. The oven need not be kept at bread heat after the plums have begun to cook thoroughly; the time of baking is usually a week, but largely depends upon the heat and care given them. The plums to be used are not the damsons, but the common blue variety, ripening early in September.

MATILDA JOSLYN GAGE.

Quince Preserve.

Put into a porcelain kettle two pounds granulated sugar with one pint cold water; when nearly boiling stir in the white of an egg beaten, as it boils up; remove the scum, and let it boil till no more scum rises; pare and quarter the quinces, and boil in clear water till tender; then pour them into the syrup and boil them all together a short time, allowing three-quarters of a pound of sugar to a pound of fruit.

MRS. ALICE M. SOUTHWICK.

Quince Preserve.

Ten pounds quinces, seven pounds pears, five pounds sugar, two quarts water; cook four and one-half hours.

JANE HOSMER.

Spiced Crabapples.

Take five pounds apples, four pounds sugar, one pint vinegar, two tablespoonfuls cloves and two tablespoonfuls cinnamon. Core the apples and use whole spices. Boil one hour.

L. CLEMENTINE GATES.

Spiced Grapes.

Eight pounds grapes, five pounds sugar, one cup vinegar, a large tablespoonful of cinnamon and half a tablespoonful of cloves. Remove the pulp from the skin, cook, and sift out the seeds. Then add skins, sugar, vinegar and spice; cook till the skins are soft. ZILPHA H. SPOONER.

Tropic Delicacy.

Slice four bananas, sprinkle profusely with powdered sugar, and squeeze the juice of two lemons over them. Put on ice till served. MRS. LILLIE DEVEREUX BLAKE.

Grape Jelly.

Boil grapes till tender, putting in just enough water to keep from burning until some juice can be pressed out. Then strain and measure the juice; boil twenty minutes, during which time heat the same amount of granulated sugar, measure for measure. Skim the juice and stir the sugar frequently; then add the sugar to the juice; stir constantly until the sugar melts; remove from the fire before it comes to the boiling point. Pour immediately into the glasses, which will not break if set on a wet cloth.

CLARA BERWICK COLBY.

COOKING FOR AND CARE OF INVALIDS.

Suggestions in the Care of Invalids.

When a tray is prepared for an invalid, everything should be very clean and neatly arranged. Drinks should never be slopped over into the saucer; the butter should be in a small plate by itself. It is well to have the milk in a little pitcher and the sugar in a tiny bowl or cup. If only a bowl of gruel is to be offered, the bowl should be the prettiest in the house, the tray covered with a napkin, and not too much carried up at once. To see a large quantity of food is often enough to take away the appetite of an invalid entirely. Do not talk about what is to be prepared for an invalid in the sick-room. Let the meal be unexpected; it will be eaten with more relish. Never let any food stand in the chamber; remove it at once after each meal. Do not let it remain on the supposition that the invalid will perhaps take a little more after a while; it will be very certain not to be used. The same may be said about anything else used in the room. Remove it at once, and allow nothing to remain to litter up the room or create an odor.

Great attention should be paid to ventilation, as to be obliged to eat in an ill-smelling apartment would revolt the stomach of a strong and healthy person. Air thoroughly bed clothing and room. Bathe the face and hands gently before a meal. Cleanliness and fresh air will do much to improve the appetite. S. Adelaide Hall, M. D.

Important Rules.

1ST. Avoid the use of coffee, opium or tobacco. Also avoid the use of hot bread, pancakes. dumplings, salt meat, pork, ham, sausage, fried eggs, pickles, gravy and rich pastry.

2D. Avoid the use of apples, nuts, or *any* kind of food except at meal time. Frequent eating enfeebles the digestive organs and causes dyspepsia, and various other diseases.

3D. "Bread is the staff of life." It ought to be made with "hop yeast," and kneaded until every atom of yeast comes in contact with one of flour, and then baked with the oven moderately heated until it is cooked through; after baking, it should be kept forty-eight hours at least before it is cut, and if a week old, all the better. Fresh bread is the bane of dyspeptics, and causes much difficulty.

4TH. Fresh beef, mutton, game, fish and eggs, properly cooked, may be used morning and noon. Beans, peas, rice, and most vegetables may be used as the stomach will bear. Eggs should be boiled, never fried.

5TH. Potatoes, custards, puddings, etc., may not always agree with weak stomachs, and so of fruit; they must be eaten as the stomach will bear, and always at meal time. Fruit should be ripe and fresh.

6TH. Eat slowly and chew your food thoroughly. Use with meals warm or cold drinks as will best agree with the stomach. Milk and hot water, or weak black tea, is much preferable to green tea. It is best not to drink while eating. Finish the meal and then take what drink the stomach needs. Most people drink too much; one cupful of tea is better than two.

7TH. Bathe regularly, always remembering to wipe dry and rub thoroughly. If quite feeble, do this in a warm room, at 10 A. M. Thorough friction over the entire surface of the body, daily, with a dry towel, flesh brush or hand is very beneficial. The nerves of the skin sympathize with all the internal organs, and a vast amount of relief may be obtained in this way.

8TH. Exercise regularly in the open air, and if possible in the sun, according to strength. Short walks, often repeated, are the best. Avoid the stooping position while sitting or standing at work.

9TH. Sleep in a large, well-lighted and well-ventilated room. Sunlight is indispensable to health, in animal and vegetable life. Dark rooms are fatal to invalids. Plants and animals will not mature without exposure to the sun.

10TH. Do not continually think and talk about your diseases; keeping the mind on a weakness makes one feel worse. Have some light, cheerful employment to occupy your thoughts so as to forget your diseases.

DR. VESTA D. MILLER.

Beef Tea

One pound of rump steak, free from fat, cut up in small pieces, and put in a covered enamelled saucepan with one pint of cold water. Let this stand in a cool place six or eight hours; then place over the fire and bring to a quick boil; remove from the fire as soon as it boils, season with salt. If wanted the tea in the morning, let it soak in cold water over night, then bring to a quick boil in the morning. A very nutritious and delicious beef tea.

ANNA B. TAYLOR, M. D.

My Original Beef Tea.

One pound fresh raw beef chopped fine; put into a porcelain kettle, add one teacupful of cold water, set away for an hour, if not in a hurry, then set it over the fire; stir, press, and knead it constantly until it begins to turn grey; do not let it boil. A strong iron spoon is *best*, lest you spoil a good one. Strain the juice all out, use a little tin strainer, return the clear juice to the kettle, scald a few moments—*not boil*—season with salt and pepper.

DR. ALICE M. EATON.

Raw Beef Sandwich for Invalids.

Scrape some raw beef fine, season with salt and pepper, and spread between two thin slices of slightly buttered bread; then cut in strips. S. ADELAIDE HALL, M. D.

Barberry Jelly,

To be used for an acid drink by adding water, is made by boiling a peck of barberries in a quart of water till quite soft, then straining off the juice and boiling it down quite thick and adding a pint of sugar to a pint of juice. This requires longer boiling than currants or other fruit, in order to evaporate the needed water which was added. When using the jelly for a drink, it is much nicer to pour hot water to dissolve it, and then let it cool.

DR. EMMA M. E. SANBORN.

Egg Lemonade.

White of one egg, one tablespoonful pulverized sugar, juice of one lemon, one goblet water; beat together. Very grateful in inflammation of the lungs, stomach, or bowels.

ALICE B. STOCKHAM, M. D.

Invalid's Gingerbread.

[From Mrs. Cornelius' YOUNG HOUSEKEEPERS' FRIEND.]

One pint of molasses, one cupful sugar, one teaspoonful of soda, one teaspoonful of ginger, one of salt, with flour enough to roll out very thin. It is improved by keeping a week or two. EMMA M. E. SANBORN, M. D.

Extract of Beef.

One pound from the round of beef, to be chopped fine; put this in one pint of cold water, add eight drops of strong muriatic acid, let it stand in a cool place over night. In the morning put the vessel containing it into a saucepan of hot water, and let it stand in this for two hours on the back part of the stove where it will not boil; after which it is ready for use. A cupful may be taken twice a day.

M. E. ZAKZEWSKA, M. D.

Koumiss.

Take two quarts each of fresh milk and warm water, add two tablespoonfuls of sugar and one-half a compressed yeast-cake; let this stand in a warm place till bubbles begin to rise, then bottle it in champagne-topped bottles, or those with the patent tops, as the rest will thus keep good if only a glass is used at a time. EMMA M. E. SANBORN, M. D.

Milk Porridge.

To half a pint of boiling water add two teaspoonfuls of flour wet smooth in cold water, and add salt; then add half a pint of milk, and boil up again; the proportions of milk and water can be varied. This is very palatable, and is excellent for people suffering with bowel complaints.

S. ADELAIDE HALL, M. D.

Weak Tea,

To which a little lemon juice is added, makes a cooling drink, or weak lemonade where much thirst exists as in fevers. The weaker the drink the longer one's relish remains for the same drink, which is preferable to ice water.

DR. EMMA M. E. SANBORN.

Toast-Water Lemonade.

An excellent drink for the sick is toast-water lemonade. Make toast water in the usual way, strain, and with this, instead of plain water, make the lemonade, and cool with ice. DR. LEILA G. BEDELL.

Foamed Egg.

Light nourishment for an invalid. Break an egg into a basin; if the yolk retains its globular form the egg is good. Take out the germ and beat the egg with an "American egg beater" till it is all foam; in a tumbler dissolve a lump of sugar in a tablespoonful of hot milk (or hot water); pour the foam upon this,—it ought to fill the glass,—stir, and it will be ready to drink. It is easy of digestion and nutritious.

REBECCA MOORE.

Tonics and Condiments for the Sick.

The dried skin of salt codfish, nicely toasted brown, is excellent to give tone to a weak stomach after recovering from almost any illness—especially cholera-morbus, or any gastric difficulty. A piece the size of from a ten-cent piece to a quarter in silver, chewed fine, and taken as often as agreeable. If properly toasted, it is very tender.

I find condiments often needful in the sick-room, eaten sparingly at meal-time,—such as a sauce of green or ripe tomatoes, known as CHILI SAUCE.—Take eighteen large green or ripe tomatoes, six good sized onions, six red peppers, twelve tablespoonfuls of sugar, three of salt, three cupfuls of cider vinegar; chop fine, mix and cook one hour. These simple but important recipes I have used in my practise with excellent results. The red pepper is a harmless, unirritating, stimulating condiment, and, properly used, it would, in most cases, take the place of alcoholic stimulants.

CLEMENCE S. LOZIER, M. D.

Cracked or Rolled Wheat.

In two quarts boiling water stir one pint cracked wheat, half teaspoonful salt. Use a farina boiler, or double kettle, and cook three hours without stirring. When done mould in dishes. Eat cold, with fruit sauce or cream and sugar.

Excellent in constipation or biliousness. The rolled wheat is preferable. Not being able to procure it ready prepared, one can crack wheat in an ordinary coffee-mill.

ALICE B. STOCKHAM, M. D.

Baths.

(EXTRACTS FURNISHED BY DR. CAROLINE E. HASTINGS.)

BRAN BATH.—Boil four pounds of bran in one gallon of water, strain, and add the liquor to sufficient water for a bath. Use to allay irritability of skin, and to soften it in squamous diseases.

SALT BATH.—Add rock-salt in the proportion of one pound to four gallons of water. Use as an invigorating bath, and to lessen susceptibility to cold.

SULPHUR BATH.—Twenty grains of *Sulphuret of potassium* to a gallon of water. For skin diseases and rheumatism.

MUSTARD BATH.---Add a handful of mustard to the ordinary hot bath, or a smaller quantity to a foot bath. Use when stimulating action is required.

COLD DOUCHE.---Lower patient's head, place rubber cloth under, and pour cold water from a pitcher over crown of head, the pitcher being slowly and gradually raised higher and higher, so that the water may fall with more force. Use in sunstroke, and intense cerebral congestion.

WET PACK.---Spread a comforter and several blankets on the bed, and over these a sheet wrung out of cold water. Remove all of the patient's clothing, lay him in the middle of the sheet, draw the edges of sheet over, and wrap the patient in it snugly, then draw over one side after another of blankets and comfort, and make all snug. Put cold, wet compress on forehead. Use to reduce temperature in typhoid, and to develop delayed eruption in scarlet, and other specific fevers. It develops the rash, greatly reduces the fever, quiets the pulse, renders the skin moist and comfortable, and abates the restlessness and wandering.

BLANKET BATH.—A blanket is wrung out of hot water, and wrapped around the patient. He is to be packed in three or four dry blankets, and allowed to rest quietly for thirty minutes. The surface of the body should then be well rubbed with warm towels and the patient made comfortable in bed. This is an easy means of inducing perspiration.

VAPOR BATH, IMPROVISED.—Place patient, with clothing removed, in large, cane-seated chair, and surround both completely with blankets, letting them extend to the flour, and be secured about patient's neck. Under the chair place basin of hot water, with alcoholic lamp beneath it; bring the water to a boil, and patient will soon be brought into a state of perspiration which may be carried to any desired extent. Use in uræmia, Bright's disease, and whenever diaphoresis is required.

MISCELLANEOUS.

Table of Weights and Measures.

One quart of flour	one pound.
Two cupfuls of butter	one pound.
One generous pint of liquid	one pound.
Two cupfuls of granulated sugar	one pound.
Two heaping cupfuls of powdered sugar	one pound.
One pint of finely-chopped meat, packed solidly,	one pound.
Ten eggs	one pound.
Eight tablespoonfuls	one gill.
One small wine-glass	one-half gill.
Two gills	one-half pint.
Forty drops	one teaspoonful.

Slipcoat Cheese.

The following receipt is from "Mrs. Glasse,"—the old, and I believe, only authority in cooking a century ago,—supposed to have been written by Dr. Johnson, quite above the capacity of a woman.

"To make Slipcoat Cheese. Take six quarts new milk, hot from the cow, the streakings, and put to it two spoonfuls of rennet; and when it is hard coming, lay it into the fat with a spoon, not breaking it at all; then press it with a four-pound weight, turning of it with a dry cloth once an hour, and every day shifting it into fresh grass. It will be ready to cut, if the weather be hot, in fourteen days."

LUCY GODDARD.

Hints for Preserving the Health.

COLD BATHING,—with care, immediately on rising, will often ward off consumption and many constitutional ills; and is a fine tonic against all weakness, even mental and moral. There may be occasionally constitutional affections—if reaction does not take place at once, and easily, the bath should not be taken.

TO GUARD AGAINST A THREATENED COLD. Call some one to rub briskly the back, between the shoulders, over the lungs. Handle and smell a bit of gum of camphor; don't expose yourself, especially the back, to draughts of air; eat no supper that night.

FOR SLEEPLESSNESS. Rub with the hand in downward strokes the base of the brain. A wet cloth is good.

LUCY GODDARD.

Protest Against Pepper.

Soups, and Their Value.

MARY J. SAFFORD, M. D.

I send this protest for the WOMAN SUFFRAGE COOK-BOOK, viz.:

That pepper, especially black pepper, be omitted from every receipt given. It is an abomination to the sense of every normal stomach. If one is so abnormally constituted as to desire it, it can be added *ad libitum*, and thus give those who cannot endure it, a chance to eat many a dish that would, without the pepper, be palatable and wholesome. There is nothing more grateful, especially in cold weather, and after fatigue, than well-prepared, nourishing soups; but at hotels, in boarding-houses, and often in private homes, they are made so unpalatable and injurious by pepper that no right-minded person would think of giving them to children, nor of eating them themselves. While speaking of soups, I would like to say that they ought to be a part of every well-ordered dinner. They may be made most nourishing and appetizing at a very small cost. Anyone who has seen the French and German peasantry make what seemed an enjoyable and satisfying meal on a well-made soup and black-bread, can appreciate how much it might add of nourishment and cheer to the laboring-man's dinner in America.

If there is a roast for dinner, any pieces that would not be suited, cold, for the table, may be used for soup the next day, with all the bones, juice and gravy that were left added, with a potato, turnip, carrot, onion, celery, and a light sprinkle of herbs, dry or green, if one has them—all to be chopped fine and put in with the meat remnants, and to simmer slowly from early morning until dinner-time. The remnants from a chicken or turkey dinner make a delicious soup; the nourishment and flavor is enhanced by adding the feet of the fowl, the heads and necks, all of which may be properly prepared by pouring boiling water upon them, which enables one to remove the skin as a glove from the finger. There is, I venture to say, enough good material wasted from the majority of American tables every day, to make an excellent soup for the day following. I have never been able to understand why this inexpensive and nourishing part of a good meal,—a soup, has been so generally neglected by our American housewives.

Plain Living and High Thinking.

I have formed a settled conviction that the world is fed too much. Pastries, cakes, hot bread, rich gravies, pickles, pepper sauces, salads, tea and coffee, are discarded from my "bill of fare," and I firmly believe that they will be from the recipes of the twentieth century. Entire wheat-flour bread, vegetables, fruit, fish with a little meat, and milk as the chief drink, will distil in the alembic of the digestive organs, into pure, rich, feverless blood, electric but steady nerves, and brains that can "think God's thoughts after

Him," as they have never yet been thought. This is my recipe: "Plain living and high thinking;" and this is my warning: With high living you will get exceedingly plain thinking.

Yours for stomachic rights,

FRANCES WILLARD.

Communion Wine.

Squeeze your grapes with your hands or otherwise, put in your kettle, and just let it come to a boil; then strain through a white flannel. Add to one gallon of the strained liquid fourteen pounds of granulated sugar. Slowly simmer (but not boil), and skim off all impurities that may arise. When thoroughly cool, put into a jug or bottle and cork tight. Place in a cool temperature. When used for the table, reduce by adding one and a-half or two pints of water to one pint of the syrup. MRS. A. A. MINER.

Brine for Beef or Hams.

To one hundred pounds of meat add nine pounds salt, six pounds brown sugar, three ounces saltpetre, one and a-half ounces potash, and six gallons of water made into a brine, and poured on hot. The meat should be covered with the brine six inches. MRS. SARA T. L. ROBINSON.

A Nice Pickle for Meat.

Five pints molasses, five ounces saltpetre, eight pounds rock-salt, and three gallons water. Boil till the salt is dissolved; skim, and when cold, it is ready for use. This is of sufficient quantity for one hundred pounds of meat closely packed. Care should be taken to have the meat *held under the brine* with a weight. Hams cured in this way before smoking are superior to Boston cured, as they are neither as salt or hard.

SARAH E. M. KINGSBURY.

Communion Wine.

One quart of grape jelly; two quarts of boiling water. Stir well, let it stand over night; then strain, and it is ready for use.

Hard Soap.

Six pounds of clean grease, and one-pound can of Babbitt's potash will make excellent hard soap. Melt the grease. Dissolve the potash in one quart of water. When both are nearly cool, or just milk-warm, put them together, and stir well until they become ropy, which should be within ten minutes. Then pour the mixture into a smooth box, or the dripping-pan will do. It hardens at once, and is ready for use the next day. LUCY STONE.

Soft Soap.

Twelve pounds of grease, and twelve pounds of crude potash will make a barrel of soft soap. Melt the grease. Dissolve the potash. Pour the grease hot into the barrel, and when the potash is cool, pour it into the hot grease; stir it a few minutes, then fill the barrel with hot water, stirring well from time to time, as the water is put in. It is well to stir it occasionally for a few weeks, but this is not essential. The older the soap, the better for use. LUCY STONE.

Directions for Treatment of Accidental Burns, Scalds and Cuts.

There is no accident which happens more frequently in domestic work than a burn or scald. A burn from a red-hot iron surface requires hardly any care, as the intensity of the heat forms a crust of the skin, which, if not exposed to wet or cold, will heal without further attention; a bandage of cotton is sufficient.

Scalding with hot liquids or steam may produce dangerous sores. If the feet or legs are scalded, a reclining position at once is absolutely necessary till fully healed. If the hands, arms or face are the places of injury, there is no need of being quiet, but simply the part injured needs rest. Immediately after the accident has occurred, it is best to expose the afflicted organ to stove-heat, as near as the pain it causes will permit; then apply lightly a fatty substance, such as sweet-oil, melted beef-dripping, mutton-fat, or lard,—whichever is nearest at hand; then cover it thickly with soda or saleratus, or if nothing else is at hand, baking powder, then remain still for an hour or two, with the injured surface toward the heat, until all the burning sensation ceases. Leave the crust of soda on which has formed, tie up the wound the best you can without causing pain, and leave it thus for three days. When the dressing is removed, a crust will have formed; if any small blisters appear, be sure not to prick or break them, and the whole surface will be well and natural again in three weeks, by simply renewing the cotton dressing.

Another accident may be a cut, with glass, knife, or nail. In such a case do *not* plunge the wound in cold water in order to stop bleeding, but use *hot* water, dipping the surface into it, or allowing it to run on the wound, or applying it with thick, hot, wet cloths. The heat will remove all pain, and healing will immediately set in. Even a bleeding artery will cease flowing when in hot water; if not, press a finger on the artery and hold it. Simple bruises will also be relieved by hot water.

MARIE E. ZAKRZEWSKA, M. D.

Cleansing Blankets.

Fill a tub with cold water, add one pint of soft soap reduced to a lather; dissolve two tablespoonfuls of borax after pounding fine, in a pint of hot water. Put the blankets to soak over night, and in the morning rinse thoroughly in cold water, then hang up to dry without wringing.

MRS. A. A. MINER.

Washing with Kerosene.

Shave thin one-half pound of good hard soap and dissolve in two-thirds boiler of water. Let it come to a boil, then add two and one-half tablespoonfuls of kerosene oil. Dip out some of the hot suds, in which the clothes that are stained or much soiled may be rubbed slightly before putting into the boiler. Boil briskly from ten to twenty minutes, rub in the suds-water if necessary, and rinse as usual. Clothes slightly soiled need not be rubbed at all before boiling. This is much easier than the old-fashioned method of washing. ANGELINE RICKETSON.

Washing Made Easy

Two pounds of soap reduced to a pulp in a little water; to this add ten gallons of water, one large tablespoonful of turpentine oil, and two tablespoonfuls of ammonia. Stir the mixture well, having it as hot as your hand can bear. Into this solution put the white clothes, soaking them two hours before washing, taking care to cover the tub meanwhile. Clothes washed in this way need no boiling, nor have I found the texture to be injured thereby.

MRS. M. F. WALLING.

Wash for Chamois Skins.

Use no soap, but take one teaspoonful of powdered borax to three pints of water, and wash thoroughly. Rinse well in three or four waters, and they will be very clean.

MRS. MARTHA J. WAITE.

To Wash Blankets.

Dissolve one ounce of borax in hot water, add two pails hot water (not boiling, but more than milk-warm), put in the blankets, handle and press them into the water until the water is colored with the dust from the blankets; ***don't wring***, but drain and rinse them *thoroughly* in warm water and hang out, stretching them into shape while drying.

MRS. D. B. WHITTIER.

To Cleanse Soiled Ribbons and Laces.

To one cup tepid water add one teaspoonful ammonia. Wash or squeeze the ribbons, and if water is very dirty, wash in another water prepared as above. Do not rinse, but iron while damp, and the colors will not be injured.

When cleansing laces, if you keep them *wet* until they are to be ironed, they will not shrink. Even cheap edgings, if kept in water until you wish to iron them, will look like new. It is drying and re-wetting them which causes the shrinkage. MRS. L. CLEMENTINE GATES.

To Remove Mildew.

Rub the spots well with soft soap, then cover with a mixture of soap and powdered chalk, and lay upon the grass in the sun; repeat the process until the discoloration disappears. MRS. M. F. WALLING.

To Remove Ink-Spots.

Apply a solution of oxalic acid, not so strong as to remove the color of the fabric, washing the article to remove the surplus acid. Iron-rust may be removed in the same way.

MRS. M. F. WALLING.

Wigglers.

To prevent wigglers in the rain-water, put a few minnows in. They eat the mosquito eggs, and thus keep the water clear. MARIANA T. FOLSOM.

To Destroy Ants.

Dissolve two teaspoonfuls of alum in a gallon of boiling water, and while hot, wash the shelves where ants congre.

gate. If they do not disappear, sprinkle powdered alum on the shelves, and they will be gone the next day.

MRS. M. F. WALLING.

To Preserve the Complexion,

and prevent wrinkles in the face, wash the face every night before retiring in tepid water with a little soap. The accumulation of dust and insensible perspiration being removed, the skin can perform its proper function, and thereby be preserved. MRS. E. C. CROSBY.

Disinfectants.

No. 1.

Copperas (*Sulphate of Iron*). This is a cheap and good disinfectant for many purposes. It can be obtained at any drug-store. In warm water it will soon dissolve by stirring; when put into cold water, let it stand all day, or over night. Use in about the following proportions:

To a gallon of water, add two pounds copperas. This can be used for privy-vaults, water-closets, catch-basins, cess-pools, etc. Pour into water-closet about a gallon at a time.

No. 2.

Chloride of Zinc. This is one of the best of disinfectants. It is superior to the copperas solution, but being more expensive, is not so available for use in large quantities. Prepare in proportion of chloride of zinc, one pound; water, two gallons. Throw this into kitchen sinks, house drains, cess-pools, water-closets, and the like; also use it in chamber-vessels, about the sick-room.

No. 3.

Bichloride of Mercury (*Corrosive Sublimate*). A solution consisting of one part of the bichloride of mercury to one thousand parts of water is one of the most efficient disinfectants known. It can be used for water-closets, urinals, sinks, and cess-pools, or for soaking clothing, towels, bedding, and other fabrics. Corrosive sublimate is a dangerous poison and should be carefully handled.

No. 4.

Carbolic Acid. This is an excellent disinfectant if used sufficiently strong, but a weak solution does little good. Two ounces of carbolic acid to one quart of water, for night vessels, sinks, and water-closets. One pint carbolic acid to five gallons water, for drains, sewers, and cess-pools.

No. 5.

Quicklime. Unslacked lime may be used to throw about wet places, in damp cellars, under buildings; stables and sheds should be whitewashed.

No. 6.

Chloride of Lime. This may be strewn about barns and outhouses, and thrown into cess-pools, drains and sewers. Do not use chloride of lime about the house. Other disinfectants, which are less offensive, are at the same time equally efficacious, and some even of greater value.

No. 7.

Charcoal. This is very useful to cover heaps of filth, pools, and wet places. Sometimes it is better not to disturb an old cess-pool, but, instead, cover it over with charcoal; dry earth may be similarly used, and it is almost as good.

No. 8.

For Soiled Clothing. Make a solution in the following manner: sulphate of zinc, one pound; carbolic acid, two ounces; water, four gallons. Keep a tubful of this near the sickroom, and into it place all soiled bed-linen and clothing. If clothing be subjected to a temperature of 212° F. for an hour, either by boiling or baking, it effectually destroys all germs. After all, the best disinfectant is *fire*, and, if possible, everything that has been in contact with the sick had better be burned.

No. 9.

For Air of Sick-room. Put into a saucer permanganate of potash, one-half ounce; oxalic acid, one-half ounce; water, one ounce. Mix well. In two hours add small quantity more of water. This will emit enough ozone, which is an active disinfectant, for a large room.

Fumigation. To fumigate a room, put some sulphur (brimstone), broken into pieces, in a tin vessel, and set this on a brick which is placed in a tub having a little water in the bottom. Set the sulphur afire and hasten from the room, having all windows and doors tightly closed, and all cracks well stuffed, even to the keyhole. Keep the room closed for six hours, then open and air it. Eighteen ounces of sulphur should be used for each space of one thousand cubic feet—a room ten feet square.

REV. ANNIE H. SHAW.

Whatever other uncertainties we may recognize in values and in markets, it will always pay for women who have money enough to have leisure, to interest themselves in bettering the condition of their sex. It has become honorable to-day for women, gentle or simple, to earn money. This is as it should be, but for us to deduce therefrom the supposition that women should engage in work only as they are paid for it, would be a lamentable mistake. We must have money to live, and ought to have enough to live well and comfortably; but while life has some supreme interests, money is not one of them. We must do our *devoir*, whether it brings us in wages or not.

JULIA WARD HOWE.

Preserving the Health.

(A RECIPE THOROUGHLY TESTED, AND FOUND NOT WANTING IN ANY PARTICULAR.)

Preserving the health is the most important preserving done in the family, and, unlike most other preserving, the fruits are obtainable at all seasons of the year. But simple as the process is, and few as are the necessary ingredients, a very limited number of housewives do this preserving for everyday use in the household.

A very simple recipe for preserving the health, and one that will never fail if directions are carefully followed, consists of a good-sized measure of common-sense, to which should be added in equal quantities by the guardians of the family life, independence and thoughtfulness. Mix well with fresh air and physical exercise. Then take the spices of sanitary, physiological, and hygienic knowledge, and grate down to the consistency for everyday usage, and stir briskly into the preserving pot. Add cold water enough to form a syrup of cleanliness and vitality. For extra flavoring, nothing is better than cheerfulness and good temper. Simmer well, and serve to the separate members of the family in carefully-considered dress, sufficiently warm and loose to permit the freedom of every organ, muscle and nerve, and insure perfect circulation of the blood.

Care must be taken in this preserving not to allow popular prejudice and accepted custom to spoil the delicate flavor given by the fruit of common-sense.

ANNIE JENNESS MILLER.

Hints to Housekeepers.

While living in a warm climate, I found nothing more wholesome than fruit juices for summer drinks. The fruit of every kind—grapes, peaches, plums, raspberries—were all made by boiling and straining the fruit and sweetening until a thick syrup was formed. A few spoonfuls of this made a glass of water most delightfully refreshing.

Salt can be kept from hardening in the salt-sprinkler by adding a small quantity of corn starch when they are filled for the table.

Nothing is nicer for a lady's toilet than lemon and glycerine combined for keeping the skin soft. Take an ounce of glycerine and add the juice of two lemons. After washing, pour a few drops on the wet hands, then rub, and dry with towel, and dust the face with baby powder, brushing it off with a soft brush.

In pulling off kid gloves, they should always be pulled from the wrist, turning them wrong side out, then straighten and fold smooth. Pulling the finger-tips always spoils them.

Mothers should never allow a child to sleep with its mouth open or snore. This habit can be broken up if attended to when the child is small. They should be early taught to clear the passage leading from the nasal organ to the throat, as this, if neglected, will cause catarrh, and it is cruel to let a child grow up with a snoring habit.

ELIZABETH L. SAXON.

Hints.

Put one tablespoonful of kerosine to a quart of boiled starch for colored clothes. Iron without sprinkling; this saves time, and gives the garment a flexible stiffness like new goods.

To save labor and dust, wash stairs instead of sweeping. Wring the cloth dry and rub the carpet, then wipe off the woodwork.

A little ammonia in water will clean hair-brushes instantly.

CLARA BERWICK COLBY.

The Make-shifts of Mrs. Orderly Poore.

REV. ADA C. BOWLES.

When the business of Mr. Poore required the removal of himself and family from the ample accommodations of a city residence to such as were afforded in a rather remote country home, Mrs. Poore's intimate friends assured her in very sympathetic tones that she would "sadly miss the convenience of her old home." "Miss the 'convenience,' no doubt," briskly replied the little woman, "but not 'sadly,' if you please. When Mr. Poore was describing to us the poor old house, and regretting that no other could be found in the location needed, he closed with this question, in a most doleful voice: 'Girls, what *will* your mother do without her many closets?' and May at once made answer, 'Why, take her little box of tools and make all she wants, to be sure.' And Alice and Ned chimed in with so many declarations of faith in my wonder-box that papa Poore closed his lamentations and 'hoped mamma Poore would take with her a good supply of whatever she felt would be of service for her new home.' And this I mean to do, and have selected all my packing-boxes with reference to their future use."

In a very brief time the city home was exchanged for one of very limited accommodations, but unlimited sunshine and pure air; and then began the triumphs of the little tool-box, willing hands and a contriving mind.

A Warm Place for Slippers and Overshoes

was the first thing provided by this happy combination. A waste-space between the kitchen chimney and wall was filled in the following manner: Two boxes, each two feet square, were laid upon their sides at each end of the space, with their open tops outward. Across these was laid a much larger box in the same position, with its cover turned into a shelf resting upon cleats made with a chisel from the cover of another box. The three boxes were held together by a few screws, the top covered with a cretonne, matching the walls in color. Then a somewkat full curtain of the same was made with small brass rings sewed at the bottom of its upper hem about two and a-half inches apart, through which a wire cord was drawn tightly, and fastened through screw-eyes at each end of the top, over which was spread as a finish a colored Turkish towel. By a mere touch this curtain could be drawn either way, disclosing a neat closet in five compartments, Ned and his father claiming each one of the lower boxes for slippers and overshoes, and dividing the middle space for their rubber boots. On the wide shelf above were mittens, gossamers, ladies' lighter overshoes, &c., and on the shelves above that flat-irons and other laundry and kitchen articles, while the broad top was claimed by thrifty Ann, the long-time maid-of-all-work, for her work-basket and the books and papers that furnished her mental regalement in hours of leisure.

"Now," said Ann, crossing her well-appointed and always-tidy kitchen, "if I only had

A Cool Closet

here in the pantry, what a saving of steps it would be."

"St. Ann, thy devotion to the house of Poore shall be rewarded," exclaimed the ready mistress; "hie thee to thy chamber sweeping, and when it is finished behold a 'miracle of design,' if Tennyson will allow the application."

There was a large window in the pantry, and its lower sash being raised its whole length, a box was found that could fill the entire space and project into the little yard on the north side of the house, and there be secured with screws. Against its sides the blinds were drawn, nearly concealing it, and there fastened with wire; over its top enamelled cloth to shed the rain, while shelves, neatly fitted within, covered with white paper, presented a very tempting array of such pantry stores as needed a cool, dry place. The window-sash being weighted, furnished a most convenient door, fitting closely against the front of the box, and also allowing the condition of everything to be seen when closed, and preventing freezing when left open at night. Ann was quite wild with delight over her new acccommodations, and arranged and re-arranged her stores with increasing satisfaction.

"I knew you could do anything you tried to," she said, turning her shining face to her little mistress, "ever since you made me such

A Pretty Bedroom

up in that bare garret, with just unbleached cotton and a few curtains. The pretty straw matting and rugs and pictures, and then such a good bed and Eastman chamber-set, just like Miss May's; yes, you can do anything."

"You can beat her," cried Ned, just in from school, "for you have made an East-*man* of an East-*lake*, and that brings him into our family, and that reminds me that I have a letter for you, mother, from Grandma Eastman, and I hope she's coming on a visit right away."

And surely it was so; and as a little indistinctness in the address had sent the letter first to another town, she would arrive almost before her room could be made ready.

"It is three-quarters of a mile to the station," groaned Ned, "and she will have left before I could get there if I walked, and father has gone to the branch mill with Major.

Now she'll have to squeeze through that narrow door of Mr. Huddup's 'deepo kerridge,' and it took two men to pull fat Mrs. Tuck through it last summer, the boys say."

"Run, Ned," said his busy mother, "and make a good fire in the brown room, for grandma always wants to take off her things in her own room, and feel 'settled,' as she says."

"I'm so glad the brown chamber has an open fire-place," said Alice, "for Grandma Eastman says no chamber is properly furnished without one, and her 'Prophet's Chamber' has a noble one. Do you know, Ann, that Grandma always entertains all the lecturers and preachers and reformers, and thinks she has learned just how to do it. This is her recipe: 'A cheery welcome, good cooking, a warm room, soft bed with plenty of clothing, and then let them alone till they've done their work.' Oh, I forgot a hot water jug 'to coax the blood out of their heads,' she says. Mother, where is the rubber bottle? I must see that grandma doesn't miss her 'home comfort,' as she always calls her red jug that I used to tug up stairs last winter."

"The rubber bottle leaks, and stone jugs are to go out of fashion immediately for bed-warming," said her mother, "for see, I have found something much better. Here is a

New Bed Warmer.

This maple syrup can, filled with a gallon of boiling water and placed upright in the middle of the bed in the morning, will keep it warm all day, renewed at evening, it will retain the heat all night, and furnish you with a gallon of warm water for your morning toilet. You see I have cut a little washer of rubber for the screw-cap which prevents slipping and the possibility of leakage; then I slip the can into this Canton flannel bag, so comfortable for the feet to touch. It rolls easily about the bed, and is just the thing for a sick person, since, placed upright, it lifts the clothing from tender, tired feet and surrounds them with warm air.

Patent applied for, for the benefit of the Massachusetts Suffrage Association."

"You blessed little make-shift marmie!" cried Alice, "you'll make a wardrobe for my chamber, yet."

"Look! look!" screamed Ned, who had been watching at the window. "She's coming up the hill, and old Scrabble has a tough pull."

Scrabble, without regard to the flourishes of the stubby whip and jerking of the patched reins, with outstretched neck and many contortions of his rusty black gauntness, was working his slow way up the steep pitch to the door yard, and the weatherbeaten face of his master, wrinkled all over with smiles at the joy he could see shining in the faces that came out for the greeting.

"I know'd ye'd be glad to see her," he chuckled, "she's the right kind; she didn't say she never *could* git inter this narrer door, nor holler when Scrabble stumbled comin' down the big hill. Sich old ladies is skerce, you bet."

Such a welcome, finished off in the brown chamber with the blaze of Ned's honestly built fire shining over the brown mossy pattern of the carpet with its glimpses of partridge vines with their red berries, the dead-leaf brown of the walls relieved by cheerful pictures, overhung with bunches of autumn leaves full of Indian summer glow; the curtains of soft Canton, with their crimson borders and bows of ribbon. "How beautiful!" said Grandma, from her low easy chair before the fire; "and you did it yourselves. Why, Ned, where did you learn how to kalsomine the ceiling and walls so well?"

"For full directions see outside of the package, brush for fifty cents at Houghton's, and a marmee who never gets stuck to help a fellow; and she did all the painting herself and made the corner shelves; and Alice painted the lambrequins for them, and May made the bureau fixings; and I think its just swell," said Ned, in one breath.

"It's just lovely, and all looks so warm and cosy. But

what are these boxes which look just as if they grew into their places, they match the rest so well?"

"These under the windows," said May, "mother calls her

Make-shift Presses.

She got a carpenter to put on the castors and the hinges. You see the covers are larger than the boxes. She lined them inside with white paper, and tacked outside (straight as she knows just how to do it, without a tack showing) this old-gold Canton flannel; then she cushioned the tops with excelsior and cotton batting, and the crimson flannel for the covers is like the borders in the curtains, you see, and they do look as if they belonged there. One holds sheets and pillow-slips, and the other blankets and spreads." "But, grandma, look at this box, and see how deep and long it is," cried Ned, perching upon its lid. "I feel sober every time I look at it, for I have to keep it filled, and just in that blessed marmee's way, too. Left-hand hard wood, next pine, next kindlings, lastly shavings, rolled in newspapers. Grandma, this is

A Model Wood Box."

"The delight of grandmothers and the terror of lazy boys," laughed May; "and there goes the supper-bell, and I hear father calling." The next day brought the village dressmaker, and, as Mrs. Poore was very busy on her arrival, Ann was directed to take her into the sewing-room, where she would find the girls and her work waiting.

It was a tiny room, every inch of which was utilized. As it contained neither closets nor shelves when selected for its present purpose, Mrs. Poore had made use of an old set of book shelves, which she had painted, together with the whole room, a soft, neutral tint.

For these shelves she had ordered from a paper box factory two dozens of strong boxes of uniform size and color, two exactly fitting upon each shelf. Red bordered labels on each were marked in plain Roman letters with their contents, such as "Linings," "Patterns," "White Cotton," "Braids," "Black Velvet," "Colored Velvet," "Black Silk," "Colored Silk," etc. While one marked "Old Linen," and another "Old Flannel" were considered as properly belonging to hospital stores, since in them could be found roll bandages, lint, plasters, etc. At the opposite end of the room a wide shelf, upon hinges, for cutting, under which stood the sewing machine, and above this a small cabinet of walnut, the dozen drawers of which were filled with thread, silks, needles and pins, and all other things needful for sewing. In every available corner a shelf was fitted for a work-basket or box, and under two stood the baskets for mending. Low chairs, hassocks and lapboards, ornamental boards for scissors of various sizes, left Miss Cutter little to ask for, and she declared it was the only place where she worked that she did not wait for things to be hunted up. "This is what I call a

Time-Saving Sewing-Room,"

said the brisk little body, as she spread out her "Newest fashions." By the time the new dresses were finished there were three convenient wardrobes in which to hang them, made after the following manner, to supplement the small closets of the chambers: Five feet from the floor strips of pine, about six inches wide and one inch thick, were screwed against the walls in the corners, upon the tops of which rested shelves fitting into the corners and having the fronts rounded slightly, and furnished with curved brass rods which supported curtains falling in soft folds to the floor. The hooks for hanging were screwed to the lower part of the wide cleats, and the shelf served to hold some of the pretty ornaments that brightened the chamber.

Slipper and Bonnet Ottamons

were made from boxes having their covers a trifle enlarged

and fastened with hinges. That they might move easily, cast-off bureau knobs were screwed upon the bottom for feet, and many "odds and ends" of upholstery were used in the coverings. The family motto of "A Place for Everything," was often difficult of application, yet Mr. Poore almost regretted the prospect of a new house, and grandmother Eastman declared "if it were not for the coming suffrage campaign, and the demands upon the 'Prophet's chamber,' she would close her own house and come again for the winter, 'Over the hills to the Poore house.'"

SUPPLEMENTARY.

[Too Late for Classification.]

White Soup.

Peal four medium sized potatoes, and let them stand in cold water one-half hour before boiling; boil in clear water until very soft; then mash and sift through a sieve; have ready in the milk boiler one quart of boiling milk, in which stir the sifted potatoes as you would flour, adding salt, a piece of butter the size of an egg, and extract of celery one teaspoonful. Let it boil one minute, and it is ready for the table, a most delicious soup. Mrs. M. H. Hunt.

Roast Beef Dinner.

Prepare the potatoes by first washing, then pare, and lay in cold water; when ready to boil, put them in a kettle of cold water over a hot fire, and as soon as they are fairly soft drain off every drop of water, set the kettle on the back of the stove, take off the cover and throw a thin napkin lightly over them, till ready to serve. One or two other vegetables will be sufficient. An Indian meal dumpling or pudding, or plain boiled rice, or, what is more fashionable, a dish of macaroni, is always in order. When it is about time to put the beef in the oven, set a dripping pan or a spider on the hottest part of the top of the stove, and if the outside of the meat is not covered with fat, grease the pan lightly;

then lay in the meat and sear it all over, turning every side down. If this is done properly, the juices will be kept in the meat, making it luscious. Now take a clean dripping pan, and after well salting and dredging the meat, put it in a moderately hot oven, and while baking, open the oven door occasionally, for a few seconds, to keep the air good. The rule for time to bake is fifteen minutes for each pound of meat; but this must be modified, according as a rare or well-done roast is required. Some acid fruit, such as currants, cranberries, or wild grapes, either stewed or in jelly, is an added luxury. No condiments are wholesome. Desserts are objectionable. One course only is more conducive to health. ABBY KELLEY FOSTER.

Jobe's Turkey.

Use two rounds of veal steak, stitch the edges together except an opening to receive the dressing. Make a dressing of bread crumbs moistened with milk, and seasoned with butter, pepper, salt, chopped onion, and parsley or sage. Stuff the "turkey," stitching up the opening, and dredge with flour. Put a little butter and water in roasting pan to baste with, adding more water as required. Roast in a moderate oven an hour; make gravy as for turkey.

WILLIE AND ALLEN JOBES.

Willow Brook Farm Corn Patties.

One cup boiled sweet corn shaved fine and scraped from the cob, one tablespoonful of butter and one of flour rubbed with the corn, one-half cup of cream and one egg; season with a little salt and pepper, and fry in small cakes upon a hot buttered griddle. MARY F. HOLMES.

EMINENT OPINIONS ON WOMAN SUFFRAGE.

In the administration of a State, neither a woman as a woman, nor a man as a man has any special functions, but the gifts are equally diffused in both sexes.—*Plato.*

I go for all sharing the privileges of the government who assist in bearing its burdens, by no means excluding women. —*Abraham Lincoln.*

In the progress of civilization, woman suffrage is sure to come.—*Charles Sumner.*

Justice is on the side of woman suffrage.—*William H. Seward.*

I think there will be no end to the good that will come by woman suffrage, on the elected, on elections, on government, and on woman herself.—*Chief-Justice Chase.*

Woman's suffrage is undoubtedly coming, and I for one expect a great deal of good to result from it.—*Henry Wadsworth Longfellow.*

For over forty years I have not hesitated to declare my conviction that justice and fair dealing, and the democratic principles of our government, demand equal rights and privileges of citizenship, irrespective of sex. I have not been able to see any good reasons for denying the ballot to woman.—*John G. Whittier.*

Women have quite as much interest in good government as men, and I have never heard any satisfactory reason for excluding them from the ballot-box.—*George William Curtis.*

I take it America never gave any better principle to the world than the safety of letting every human being have the power of protection in his own hands. I claim it for woman. The moment she has the ballot, I shall think the cause is won.—*Wendell Phillips.*

Those who are ruled by law should have the power to say what shall be the laws, and who the law-makers. Women are as much interested in legislation as men, and are entitled to representation.—*William Lloyd Garrison.*

To have a voice in choosing those by whom one is governed, is a means of self-protection due to every one. Under whatever conditions, and within whatever limits, men are admitted to the suffrage, there is not a shadow of justification for not admitting women under the same. —*John Stuart Mill.*

However much the giving of political power to women may disagree with our notions of propriety, we conclude that, being required by that first pre-requisite to greater happiness, the law of equal freedom, such a concession is unquestionably right and good.—*Herbert Spencer.*

Suppose, for the sake of argument, we accept the inequality of the sexes as one of nature's immutable laws; call it a fact that women are inferior to men in mind, morals and physique. Why should this settle or materially affect the subject of so-called woman's rights? Would not this very inferiority be a reason why every advantage should be given to the weaker sex, not only for its own good, but for the highest development of the race?—*Huxley.*

Any influence I may happen to have is gladly extended in favor of woman suffrage.—*Lydia Maria Child.*

Every year gives me greater faith in it, greater hope of its success, and a more earnest wish to use what influence I possess for its advancement.—*Louisa M. Alcott.*

I earnestly desire to see a more rational basis for the political future of our sex.—*Elizabeth Stuart Phelps.*

I think women are bound to seek the suffrage as a very great means of doing good.—*Frances Power Cobbe.*

If prayer and womanly influence are doing so much for God by indirect methods, how shall it be when that electric force is brought to bear through the battery of the ballot-box.—*Frances E. Willard.*

When you were weak and I was strong, I toiled for you. Now you are strong and I am weak. Because of my work for you, I ask your aid. I ask the ballot for myself and my sex. As I stood by you, I pray you stand by me and mine. —*Clara Barton, to the Soldiers.*

If the principle on which we founded our Government is true, that taxation must not be without representation, and if women hold property and are taxed, it follows that women should be represented in the State by their votes. . . . I think the State can no more afford to dispense with the votes of women in its affairs than the family.—*Harriet Beecher Stowe.*

It is difficult to choose names when the list is so long, but it is right to mention among the distinguished women who have been with this movement from the outset, the names of Mrs. Somerville, Harriet Martineau, Florence Nightingale, Mrs. Browning, Miss Anna Swanwick, Miss Cobbe, Mrs. Grote, Mrs. Ritchie (Miss Thackeray), Mary Carpenter and Mrs. Jameson.—*Millicent Garrett Fawcett.*

One principal cause of the failure of so many magnificent schemes, social, political, religious, which have followed each other age after age, has been this: that in almost every case they have ignored the rights and powers of one-half the human race, viz., women. I believe that politics will not go right, that society will not go right, that religion will not go right, that nothing human will ever go right, except in so far as woman goes right; and to make woman go right she must be put in her place, and she must have her rights.—*Charles Kingsley.*

Woman must be enfranchised. It is a mere question of time. She must be a slave, or an equal; there is no middle ground. Admit, in the slightest degree, her right to property or education, and she must have the ballot to protect the one and use the other. And there are no objections to this, except such as would equally hold against the whole theory of republican government.—*T. W. Higginson.*

In quite early life I formed the opinion that women ought to vote, because it is right, and for the best interests of the country. Years of observation and thought have strengthened the opinion.—*Bishop Bowman.*

I fully believe that the time has come when the ballot should be given to woman. Both her intelligence and conscience would lead her to vote on the side of justice and pure morals.—*Bishop Hurst.*

I believe that the great vices in our large cities will never be conquered until the ballot is put in the hands of women. —*Bishop Simpson.*

In view of the terrible corruption of our politics, people ask, Can we maintain universal suffrage? I say no, not without the aid of women.—*Bishop Gilbert Haven.*

We need the participation of woman in the ballot-box. It is idle to fear that she will meet with disrespect or insult at the polls. Let her walk up firmly and modestly to deposit her vote, and if any one ventures to molest her, the crowd will swallow him up as the whale swallowed Jonah. —*Henry Ward Beecher.*

In re woman suffrage, I know of many prejudices against it, but nothing which deserves to be called a reason. The reasons are all on the other side.—*Professor Borden P. Browne, Boston University.*

I believe that the enfranchisement of woman would be a direct benefit both to woman herself and to the State.—*Rev. Charles F. Thwing.*

I believe that the admission of woman to the suffrage is in the line of God's providence, and that it is approved by the spirit of the Bible and the experience of history.—*Rev. William Burnet Wright.*

Why should not women vote? The essence of all republicanism is that they who feel the pressure of the law shall have a voice in its enactment.—*Rev. John Pierpont.*

I have not found a respectable reason why women should not vote, although I have read almost everything that has been written on the subject, on both sides.—*M. J. Savage.*

Voting would increase the intelligence of women, and be a powerful stimulus to female education. It would enable women to protect their own industrial, social, moral and educational rights. . . . Woman's vote would be to the vices in our great cities what the lightning is to the oak. . . . I believe that this reform is coming, and that it will come to stay.—*Joseph Cook.*

I leave it to others to speak of suffrage as a right or a privilege; I speak of it as a duty. . . . What right have you women to leave all this work of caring for the country with men? Is it not your country as well as theirs? Are not your children to live in it after you are gone? And are you not bound to contribute whatever faculty God has given you to make it and keep it a pure, safe and happy land?—*James Freeman Clarke.*

It is very cheap wit that finds it so droll that a woman should vote. . . . If the wants, the passions, the vices, are allowed a full vote, through the hands of a half-brutal, intemperate population, I think it but fair that the virtues, the aspirations, should be allowed a full voice as an offset, through the purest of the people.—*Ralph Waldo Emerson.*

The correct principle is that women are not only justified, but exhibit the most exalted virtue, when they enter on the concerns of their country, of humanity, and of their God.—*John Quincy Adams.*

All I have done for negro suffrage, I will do for woman suffrage.—*Henry Wilson.*

I am highly gratified with the late demonstration in the Senate, on the question of female suffrage.—*Hon. George W. Julian.*

When we seriously attempt to show that a woman who pays taxes ought not to have a voice in the manner in which the taxes are expended, that a woman whose property and liberty and person are controlled by the laws should have no voice in framing those laws, it is not easy. If women are fit to rule in monarchies, it is difficult to say why they are not qualified to vote in a republic. — *Hon. H. B. Anthony, R. I.*

Laugh as we may, put it aside as a jest if we will, keep it out of Congress or political campaigns, still, the woman question is rising in our horizon larger than the size of a man's hand; and some solution, ere long, that question must find.—*James A. Garfield.*

Marian
Brubaker

THOMAS D. COOK,

Wedding Receptions and other Parties
will receive Special Attention.

LADIES' LUNCH.

23 to 31 Avon Street - - - - BOSTON.

"WHICH WAY?"

That every article used for the same purpose is of equal value, no one admits. As in all natural productions, so in the manufactured, all have their degrees of merit, and Soap is as conspicuous in its variety as anything else. But in this, as the demand for the ***Pure and Reliable*** increases, so does the value of the productions of CURTIS DAVIS & CO., especially their ***"Welcome and Unequalled Extra"*** brands, being practically recommended by other manufacturers, who imitate them in every conceivable way. But, while this compliment may be appreciated, what can be said in favor of such competition, or the character of such goods and their makers?

No one should be deceived, as the original has the name of CURTIS DAVIS, in full, either on the bar or wrapper, and it is their purpose to let their reputation stand on this class of goods.